KE

This index is designed to find references in the Code. Having worked in the electrical field for over 39 years as an apprentice, journeyman, master, electrical inspector, instructor, and author, I realize at times how difficult it is to "find it in the Code."

The Code is laid out in Articles and Chapters starting with Article 90, the introduction. Chapter 1 Article 100 is definitions, Article 110 is requirements for electrical installations. Chapter 2 is wiring and protection with Article 200 use and identification of grounded conductors, Article 210 branch circuits, 215 feeders, 220 calculations and so on. Chapters 2, 3 and 4 are the "meat" of the Code for the electrician in his daily work. Chapters 5, 6, 7 and 8 are for special applications such as hazardous locations, elevators, pools, mobile homes, studios, high-voltage, signs, health care facilities, etc. Chapter 9 is a special chapter with tables and examples.

The first part of the Article is for under 600 volts, the latter part of the Article is for over 600 volts. All the rules are not in the Electrical Code. Example: What is the minimum height of a ceiling paddle fan in a residence? You would find that information in the Mechanical Code.

110-3b in the Code is very important as it states, "all electrical equipment shall be installed as listed." Which means the directions of installation that come with the equipment must be followed.

Copyright © 1995 by Tom Henry. All rights reserved. No part of this publication may be reproduced in any form or by any means: electronic, mechanical, photo-copying, audio or video recording, scanning, or otherwise without prior written permission of the copyright holder.

While every precaution has been taken in the preparation of this book, the author and publisher assumes no responsibility for errors or omissions. Neither is any liability assumed from the use of the information contained herein.

National Electrical Code ® and NEC ® are Registered Trademarks of the National Fire Protection Association, Inc., Quincy, MA.

3rd Printing   ISBN   0-945495-46-3

KEY WORD INDEX pg - 1

-A-

A/C rated-load current 440-2 p.458
Abandoned outlets 354-7 page 287
Above floor level, garages 511-3a p.539
Above ground storage tanks T.515-2 p.552
Abrasion 347-12 page 269
Abrasives 110-12c p.40
Absorbent materials 300-6c page 166
Absorbing regenerated power 620-91a p.721
AC - DC in same enclosure 300-3c1 p.160
AC - DC in same enclosure 725-26a p.823
AC - DC, snap switch inductive 380-14b2 p.324
AC - DC, tapped 210-10 page 58
AC cable, definition 333-1 p.242
AC circuit less than 50v grounded 250-5a p.123
AC resistance, Table 9 p.889
AC snap switch 380-14a page 324
Accelerate their loads 440-54b p.468
Acceleration, motor 430-34 FPN p.431
Access to elect. equipment 760-5 p.834
Access to elect. equipment 770-7 p.845
Access to elect. equipment 725-5 p.820
Access to elect. equipment 800-5 p.853
Access to elect. equipment 820-5 p.871
Access, electric equipment 110-16 p.41
Accessible, attics 333-12 p.244
Accessible, boxes 370-29 p.310
Accessible, electrode connection 250-112 p.152
Accessible, readily, disconnect A/C 440-14 p. 463
Accessible, sign transformer 600-21a page 692
Accessible, switches 6' 7" 380-8a p.322
Accessible, to pedestrians 600-5a p.690
Accessory buildings 210-8a2 p.57
Acetone 500-3a4 p.496
Achieve balance 310-4 FPN p.177
Acid or alkali type 700-12a p.793
Acrolein 500-3a2 p.496
Adapter, polarized 410-58b3 page 372
Adapters 240-54 page 117
Adapters, two-fers 520-69 p.613
Additional lighting fixtures 110-16d p.43
Adequacy 90-1b page 21
Adequate access cable tray 318-6i p.219
Adequate bonding 250-44 FPN p.133
Adequate compaction of fill 300-5f page 162
Adequately bonded, cablebus 365-2a p.300
Adjacent light source 110-16d p.43
Adjacent to basin location 210-52d p.65
Adjoining wireways, rigid joint 362-8b p.292
Adjustable speed drive 430-2 page 414
Adjustable speed drive DEF 100 p.31
Adjustable trip circuit breaker 240-6b p.110
Adobe 336-5b3 page 249
Aerial cable 342-1 p.255
Afford no protection 230-95b FPN 4 page 104
Aggregate 349-4 page 273
Agricultural buildings, wiring 547-4 p.628
Air changes per hour 511-3 ex. p.539
Air cond. branch-circuit select. 440-6a ex.1 p.461
Air conditioner cord length 440-64 p.469
Air conditioners, plug connected 440-62c p.469
Air conditioners, room 440-62c p.469
Air conditioning equipment 440-1 p.458

Air ducts 250-80 FPN page 144
Air handling spaces 300-22 p.171
Air hoses 668-31 p.748
Air moving device 424-59 FPN page 396
Air space 300-6c page 166
Air space, cabinets & cutout boxes 373-2a p.313
Airspace, concrete, masonry, tile 373-2 ex p. 313
Air voids, romex 336-4a page 248
Air, free circulation of 110-13b page 40
Air-break isolating switch 230-204a page 105
Aircraft fuel tanks 513-2c p.542
Aircraft hangars, pit 513-2a p.542
Aircraft hangars, stock rooms 513-2d p.542
Aircraft hangars, ventilated 513-2d p.542
Airport runways T.710-4b ex.5 p.806
Airport runways, cable T.300-5 p.164
Airspace for nometallic boxes 300-6c ex. p.166
Alarm threshold value 517-160b1 ex. p.598
Alarm threshold value 517-160b2 ex. p.598
Alarm, burglar 800-1 p.852
Alarm, fire systems 760-1 p.834
Alerting system, alarm 517-32c p.579
Alkali-type battery cells 480-5b p.491
Alternate source for emerg. systems 700-26 p.796
Aluminum bars, ampacity 374-6 p.319
Aluminum conductors CO/ALR 410-56b p. 370
Aluminum conductors size 310-14 p.182
Aluminum fittings & enclosures 348-1 ex. p.270
Aluminum grounding conductor 250-92a p.148
Aluminum inhibitor required 110-14 p.40
Aluminum neutral underground 230-30 ex.d p.93
Aluminum not magnetic mat. 300-20 FPN p.171
Aluminum siding, grounding 250-44e FPN p.133
Amateur transmitting station T. 810-52 p.869
Ambient temp. exceeds 86°F T.310-16 p.191
Ambient temperature 310-10(1) page 179
Ambulatory health care center DEF 517-3 p. 566
American Wire Gage (AWG) 110-6 page 39
Ammeter conductors 300-3c2 ex.3 page 160
Ammeter, anesthetizing 517-160b3 FPN p.598
Ammonia 500-3a4 p.496
Ampacities, voltage drop 310-15 FPN p. 190
Ampacity, 15 minute motor T.610-14a p.700
Ampacity, adjustment factors Note 8 page 196
Ampacity, aluminum T.310-16 p.191
Ampacity, aluminum bars 374-6 p.319
Ampacity, Appendix B p.920
Ampacity, bare cond. & insulated Note 5 p.195
Ampacity, bare conductor Note 5 p. 195
Ampacity, bare copper bars 374-6 p.319
Ampacity, bus bar 374-6 p.319
Ampacity, cables type W & G T.400-5b p.346
Ampacity, capacitor conductors 460-8a p.487
Ampacity, Class 1 conductors 725-16 p.823
Ampacity, conduit in free air T.B-310-1 p.920
Ampacity, covered conductor T. B-310-4 p.923
Ampacity, definition 100 page 24
Ampacity, extra-hard usage cords T.520-44 p.605
Ampacity, feeder minimum 215-2a p.67
Ampacity, ferromagnetic envelope 426-40 p.407
Ampacity, fixture wires 402-5 page 357
Ampacity, flex. cords & cables T.400-5a p.345
Ampacity, free air T.310-17 p.192
Ampacity, generators 445-5 p.470

KEY WORD INDEX

Ampacity, hoists & cranes T.610-14a p.700
Ampacity, knob and tube wiring 324-5b p.230
Ampacity, low-volt. rec. veh. 551-10e1 p.649
Ampacity, lowest value 310-15c page 190
Ampacity, monorail hoist T.610-14a p.700
Ampacity, motor feeder cond. 430-24 p.426
Ampacity, motor resistors T.430-23c p.426
Ampacity, neutral conductor solar 690-62 p.782
Ampacity, NM-NMB cable 60°C 336-30b p.251
Ampacity, phase converter 455-6 p.484
Ampacity, power resistors T.430-29 p.428
Ampacity, resistance welders 630-31 p.728
Ampacity, romex 60°C 336-30b page 251
Ampacity, selection 310-15c page 190
Ampacity, solar neutral 690-62 p.782
Ampacity, solar system 690-8a p.778
Ampacity, stage lights 520-42 p.604
Ampacity, temperature rating 110-14c p.41
Ampacity, transformer tie 450-6a2 p.477
Ampacity, tray cable 340-7 page 255
Ampacity, UF cable 60°C 339-5 p.254
Ampacity, welders AC and DC 630-11 p.726
Ampacity, X-Ray equip. 517-73b FPN p.595
Ampacity, X-ray momentary 660-6b p.737
Ampacity. armored cable 333-20 p.244
Ampere, fraction .5 and larger p.897
Amplifier output circuits 640-5 p.731
Amplifiers & rectifiers, type 640-11 p.732
Amusement rides clearance 525-12 p.614
An appliance, infrared lamps 422-15c page 382
Ancillary areas, back stage 520-47 p.606
Anchoring, FCC cable 328-16 p.236
Anesthetics, flammable Class I 517-60a2 p.590
Anesthetizing location, definition 517-3 p.566
Anesthetizing location, recept. 517-61a5 p.591
Anesthetizing, ammeter 517-160b3 FPN p.598
Angle pull, boxes 370-28a2 p.310
Animal behavior or prod. 547-8b FPN p.630
Animal confinement areas 547-8b p.630
Animal excrement, corrosive 547-1b p.628
Anode-positive plate tube 640-10b p.732
Anodizing, electroplating 669-1 p.748
Antenna box barrier 810-18c p.867
Antenna conductors, clearance 810-54 p.869
Antenna grd. conductor #14 820-40a3 p.874
Antenna systems, material 810-11 ex. p.866
Antenna, receiving station grd. 810-21h p.869
Antennas, indoor lead-ins 810-18c p.867
Antennas, radio-TV 225-19b p.88
Appendix B p.918
Appendix C p. 936
Appliance, fixed definition 550-2 p.631
Appliance, nonmotor overcur. 422-28e p. 386
Appliance, outlet 210-50c p.63
Appliance, portable definition 550-2 p.631
Appliance, stationary definition 550-2 p.631
Appliances 422 page 379
Appliances in transit 550-9a p.637
Appliances, battery powered 517-64e p.594
Appliances, continuous load 422-4a ex.2 p. 380
Appliances, cord polarity 422-23 page 384
Appliances, disconnect 422-21a page 383
Appliances, fastened in place 50% 210-23a p.61
Appliances, flatirons-smoothing 422-13 p.382
Appliances, live parts exposed 422-2 ex. p.379
Appliances, nameplate marking 422-30a p. 387
Appliances, no nameplate O.C.P. 422-28e p.386
Appliances, overcurrent protection 422-5 p. 380
Appliances, resistance-heating 422-28f p. 386
Appliances, signal heat 422-12 page 382
Appliances, unit switches 422-25 page 385
Appliances, water heaters b.c. 422-14b p.382
Applicable derived system 250-74 ex.4 p.141
Approved 110-2 page 38
Approved bushing fittings Class I 501-4b p.505
Approved power outlet 305-4c page 174
Approved stapling, heat cables 424-41f p.394
Aquarium, shall be grounded 250-45c p.134
Arc lamps, stage portable 520-61 p.611
Arc or thermal effect DEF 100 p.31
Arc projectors, conductor size 540-13 p.625
Arc welders 630-1 p.725
Arcing across discontinuities 501-8b FPN2 p.513
Arcing effects, equipment 110-3a6 page 38
Arcing equipment Class I 511-7a p.541
Arcing or suddenly moving parts 240-41 p.116
Arcing parts, electrical equipment 110-18 p.44
Area circular mils Table 8 p.888
Area sq. inch, insulated conduct. Table 5 p. 883
Area square inch, bare conduct. Table 8 p.888
Area square inch, compact alum. Table 5a p.887
Area square inch, conduit Table 4 p.880
Armature lead, DC generator 445-4e p.470
Armature shunt resistor 430-29 p.427
Armored cable, conductors 333-20 ex. p.244
Armored cable, construction 333-19 page 244
Armored cable, exposed work 333-11 ex.1 p. 244
Armored cable, floor joists 333-12a page 244
Armored cable, uses permitted 333-3 p.242
Armories, place of assembly 518-2 p.599
Arms & stems, tubing threads 410-38a p.368
Arms or stems, fixtures 410-28c page 365
Arranged to drain, raceways 230-53 p.97
Arranged to drain. raceway on bldg. 225-22 p.88
Array, solar system definition 690-2 p.776
Articles 511 through 517 510-1 p.538
Artificial illumination 700-1 FPN3 p.789
As low as practicable, capacitor 460-8b2 p.487
Askarel-insulated, trans. 450-25 p.480
ASME rated & stamp. vessel 422-28f ex.3 p. 386
ASME rated & stamped vessel 424-72a p.397
Asphalt, heating cables 426-20c1 page 405
Assembly halls definition 518-2 p.599
Assembly, 100 or more persons 518-1 p.598
Associated apparatus 504-2 FPN2 p.531
Asymmetrical fault current 710-21d4 p.811
Atomization charging 516-4 p.562
Atomizing heads, electrostat. equip. 516-4b p.562
Attach. plug, first-make,last-break 250-99a p.151
Attachment plug, X-Ray 517-72c p.597
Attachment plug, nonpolarized 422-23 ex p.384
Attachment plugs, cords 400-7b page 347
Attachment plugs, polarized 200-10b ex. p.51
Attachment point, service head 230-54c ex. p.97
Attachment to crossarms 800-10a2 p.853
Attendant electrical trailing cable 90-2b2 p.21
Attended outlet 410-57b ex p.371
Attended self-service stations 514-5b p.549

KEY WORD INDEX pg - 3

Attic entrance, AC cable 333-12a page 244
Attic equipment 210-63 p.67
Audible & visual indicators 517-19e ex. p.573
Audible & visual signal devices 700-7a p.790
Audio-program signals conductors 640-5 p.731
Auditoriums, place of assembly 518-2 p.599
Augmented, grounding electrode 250-84 p.146
Authority having jurisdiction 555-8 p.688
Authority having jurisdiction 90-4 page 22
Automatic closing fire dampers 450-45e page 483
Automatic means, capacitors 460-28b p.489
Automatic sprinkler, transfor. 450-43a ex. p.482
Automatically deenergize 501-8a p.512
Automatically started, motors 430-35b p.431
Automatically starting, gen. 700-12b1 p.793
Automatically transfering fuel 700-12b3 p.793
Automotive charging equipment Art. 625 p.721
Automotive diagnostic equip. GFCI 511-10 p.541
Automotive spray booths 516-3d ex.2 p.562
Automotive vehicles, not in Code 90-2b1 p.21
Autotransformer, circuit derived 210-9 p. 58
Autotransformer, cont. neutral 450-5a4 p.476
Autotransformer, dimmers 520-25c p.603
Autotransformer, fixtures 410-78 p.375
Autotransformer, motor starter 430-82b p.441
Autotransformer, phase current 450-5 FPN p.476
Autotransformers, rec. vehicle 551-20e p.651
Aux. gutters, sound equip. 640-4 ex.a p.730
Auxiliary equip. elect.-dis. lamp 410-54a p.370
Auxiliary gutters, ampacity 374-6 p.319
Auxiliary gutters, bus bar spacing 374-7 p.319
Auxiliary gutters, extend beyond 374-2 p. 318
Auxiliary gutters, nonmetallic 374-9e2 p.320
Auxiliary gutters, sheet metal 374-9e1 p.320
Auxiliary gutters, support 374-3 page 318
Auxiliary nonelectric connections 668-31 p.748
Auxiliary rectifiers, induction heat 665-24 p.741
Available current interrupting rating 110-9 p.39
Avoid heating, by induction 300-20a page 171
AWG size, American Wire Gauge 310-11 p.180

-B-

Back-fed panelboard devices 384-16f p.329
Backfill, damage to raceways 300-5f p.162
Baffle plates next to heaters 424-59 FPN p.396
Balancer sets, generators 445-4d p.470
Balancing branch circuits 220-4d page 71
Balconies, access working space 110-33b p.46
Balconies, clearance 230-9 p.91
Ballast compartment, THW 410-31 page 367
Ballast marking, fixture 410-35a page 367
Ballast, autotransformers 410-78 page 375
Ballast, primary leads 300-3c2 ex.2 p.160
Ballasts, amp rating 210-22b page 61
Ballasts, conductor within 3 410-66b p.373
Ballasts, exposed 410-76a page 375
Ballasts, thermal protection 410-73e page 374
Barbed wire 110-31 p.45
Bare conductor, ampacity Note 5 p.195
Bare conductor, area sq.in. Table 8 p.888
Bare conductor, concrete encased 250-81c p.145
Bare conductors Note 5 p.195
Bare conductors 30 volts or less 411-5c p.379

Bare copper concentric, type USE 338-1b p.252
Bare copper cond. underground 230-30 ex.c p.93
Bare live parts, working clearance T.110-16 p.42
Bare neutral alum. underground 230-30 ex.d p.93
Barns, wet & damp location 410-4a p. 358
Barriers, cabinets wiring space 373-11d p.317
Barriers, panelboard bus bars 384-3a p.325
Baseboard heaters, recepts. 424-9 FPN p. 388
Basin, bathroom 210-52d page 65
Basin, receptacle adjacent 210-52d p.65
Bathroom, branch circuit 210-52d p.65
Bathroom, definition DEF 100 page 25
Bathrooms O.C.P. 240-24e p.115
Bathrooms, GFCI 210-8a1 page 57
Bathtub and shower space recpt. 410-57c p.371
Bathtub rim, light fixture 410-4d page 359
Bathtubs, ceiling fans 410-4d p.359
Bathtubs, light fixtures 410-4d p.359
Batteries 480 p.490
Batteries, cell explosion 480-9b p.492
Batteries, cells in jars 480-5b p.491
Batteries, cells per tray 480-5b p.491
Batteries, emergency lts. 1 1/2hrs 700-12a p.793
Batteries, racks and trays 480-7a p.491
Batteries, stand-by power 87 1/2% 701-11a p.798
Batteries, vented cells flame arrest. 480-9a p.492
Batteries, volts per cell 480-2 p.490
Battery interconnections 690-74 p.784
Battery pack units emerg.power 700-12e p.794
Battery powered appliances 517-64e p.594
Beating rain, receptacles 410-57a page 371
Below finished grade 300-5d page 162
Bending machine 348-9 ex. page 271
Bending radius, armored cable 333-8 p.243
Bending radius, conductors 110-3a3 p.38
Bending radius, conduit T.346-10 p.265
Bending radius, flex. metal T. 349-20a p.274
Bending radius, gas cable T.325-11 p.232
Bending radius, hi-voltage cable 300-34 p.173
Bending radius, lead-covered 300-34 p.173
Bending radius, metal clad cable 334-11 p.246
Bending radius, MI cable 330-13 p.239
Bending radius, nonshielded cables 300-34 p.173
Bending radius, romex 336-16 p.250
Bending radius, shielded cables 300-34 p.173
Bends, cable 338-6 p.252
Bends, number in one run 346-11 page 265
Bends, rigid conduit 346-10 page 265
Bent by hand ENT 331-1 p.240
Bent, parts 110-12c p.40
Beverage dispensers cord connected 440-13 p.463
Bi-plane examinations 517-73a2 p.595
Bias supplies, induction heating 665-24 p.741
Binding screws or studs 110-14a ex. p.40
Blanketing effect, dust 500-3b3 FPN 2 p.497
Bleeder resistor, induction heat 665-24 p.741
Block, definition 800-30a FPN1 p.855
Blocking diode, definition 690-2 p.776
Boatyards & marinas 555 p.686
Boiler, stamped vessel 424-72a page 397
Boilers, electrode type 424-82 page 399
Boilers, elements subdivided 424-72b p.397
Boilers, resistance type 424-70 page 397
Bolts, screws, or rivets 370-23c page 307

KEY WORD INDEX

Bolts, screws, or rivets 410-16c page 363
Bonding all piping 250-80b FPN p.144
Bonding chassis, park trailer 552-57b p. 683
Bonding electrodes, comm. 800-40d p.858
Bonding grid, swimming pools 680-22b p. 764
Bonding jumpers main & equip. 250-79 p. 142
Bonding jumpers size T.250-94 250-79d p.150
Bonding metal air ducts 250-80b FPN p.144
Bonding to assure continuity 250-70 p.139
Bonding well casings 250-43(l) p.133
Bonding, Class I, Div. I & 2 501-16a p.516
Bonding, structural steel 250-80c p. 144
Border light 520-2 DEF p.600
Border lights, stage 520-41 p.604
Bored holes in wood 300-4a1 page 160
Bottom shield, FCC cable 328-2 page 234
Bowling lanes, place of assembly 518-2 p.599
Box, opposite wall 370-28a2 page 310
Box, plugging 530-2 p.617
Box, scatter 530-2 p.616
Boxes 370 page 302
Boxes, 100 cubic inch 370-40b page 311
Boxes, accessibility 370-29 p.310
Boxes, angle or U pulls 370-28a2 page 310
Boxes, barriers 370-28d p.310
Boxes, bushings 370-42 page 311
Boxes, cable clamps 370-16b2 page 304
Boxes, cable entry/exit over 600v 370-71b p.312
Boxes, cast metal 370-40b page 311
Boxes, clamp fill 370-16b2 p.304
Boxes, combustible walls 370-20 p.306
Boxes, conductors racked up 370-28b page 310
Boxes, corrosion-resistant 370-40a page 311
Boxes, covers suitable for condit. 370-28c p.310
Boxes, Danger High-voltage 370-72e p.313
Boxes, device or equipment fill 370-16b4 p.305
Boxes, domed covers 370-16a page 303
Boxes, equip. grounding conductor 370-16b5 p.305
Boxes, emergency circuits 700-9a p.791
Boxes, enclosing flush devices 370-19 page 306
Boxes, extension rings 370-16a page 303
Boxes, fixture studs 370-16b3 page 305
Boxes, floor listed for application 370-27b p.309
Boxes, grounding with tapped hole 370-40d p.311
Boxes, hi-voltage marking 370-72e p.313
Boxes, hickeys 370-16b3 p.305
Boxes, in wall or ceiling 370-20 page 306
Boxes, internal depth 370-24 p.309
Boxes, malleable iron 370-40b page 311
Boxes, marking 370-44 page 312
Boxes, metal faceplates grd. 370-25a p.309
Boxes, metal fixture canopies 370-25a FPN p.309
Boxes, NM cable extend 1/4" 370-17c p.306
Boxes, nonmetallic 370-17c page 306
Boxes, nonmetallic 370-3 page 302
Boxes, nonmetallic 370-43 page 311
Boxes, not required 320-16 page 228
Boxes, number of wires 370-16a1 page 304
Boxes, organic coatings raintight 300-6a p.165
Boxes, over 100 cubic inch 370-40c p.311
Boxes, over 6 feet 370-28b page 310
Boxes, over 600 volts angle pull 370-71b p. 312
Boxes, over 600 volts straight pull 370-71a p.312
Boxes, paddle fans 370-27c page 309

Boxes, pendant supports 370-23g page 308
Boxes, permanent barriers 370-28d page 310
Boxes, plaster rings 370-16a page 303
Boxes, porcelain covers 370-41 page 311
Boxes, pull and junction 370-28a page 310
Boxes, round 370-2 p.302
Boxes, shall have a cover 370-25 p.309
Boxes, shallow 1/2" internal depth 370-24 p.309
Boxes, show windows 370-27b ex. page 309
Boxes, straight pulls 370-28a1 page 310
Boxes, straight pulls over 600v 370-71a p.312
Boxes, support fittings fill 370-16b3 p.305
Boxes, support metal braces 370-23b2 p.307
Boxes, support of ceiling fans 422-18 page 383
Boxes, supported by single conduit 370-23g2 p.309
Boxes, thickness of metal 370-40b page 311
Boxes, unused openings 370-18 page 306
Boxes, volume per wire T. 370-16b page 305
Boxes, wet location 370-15 p.303
Boxes, where required 300-15a page 168
Boxes, wood brace support 370-23b2 page 307
Boxes, yoke or strap 370-16b4 page 305
Boxless device 300-15b ex.4 p.168
Boxless device NM and NMC 336-21 p.250
Braces or guys, service mast 230-28 page 93
Bracket wiring, scenery 520-63a p.611
Brackets open wiring outdoors 225-12 p.86
Brackets or cleats TC cable 340-5 page 255
Braided-covered conductors 710-5 p.807
Branch circuit, 30v or less lights 411-6 p.379
Branch circuit, air conditioner 440-1 p.458
Branch circuit, appliances 422-4 p.379
Branch circuit, balanced 220-4d p.71
Branch circuit, bathroom 210-52d p.65
Branch circuit, busway 364-12 p.298
Branch circuit, calculation 220-3 p.69
Branch circuit, classification 210-3 page 53
Branch circuit, color coding 210-5 p.54
Branch circuit, combo loads 210-22a page 61
Branch circuit, continuous 210-22c page 61
Branch circuit, cook equip. demand Note 4 p.76
Branch circuit, cord-plug 80% 210-23a page 61
Branch circuit, fixed appl. 50% 210-23a p. 61
Branch circuit, heat 424-3 p.388
Branch circuit, individual 210-3 p.53
Branch circuit, maximum load 210-22 page 61
Branch circuit, maximum voltage 210-6 p.54
Branch circuit, motors 430-22a p.424
Branch circuit, multioutlet 210-3 p.53
Branch circuit, multiwire 210-4 p.53
Branch circuit, number req. Example 1a p.897
Branch circuit, O.C.P. motors T.430-152 p.458
Branch circuit, patient bed 517-19a p.572
Branch circuit, permissible loads 210-23 p.61
Branch circuit, range neutral 210-19b ex.2 p.59
Branch circuit, ranges 210-19b page 59
Branch circuit, rating 210-3 p.53
Branch circuit, stage lights 520-41 p.604
Branch circuit, standard classification 210-3 p.53
Branch circuit, water heater 422-14b p.382
Branch circuits required 220-4 page 71
Branch circuits, 30 amp 210-23b page 62
Branch circuits, 40-50 amp 210-23c page 62
Branch circuits, data processing 645-5 p.733

KEY WORD INDEX pg - 5

Branch circuits, min. ampacity 210-19a p. 58
Branch circuits, originate 305-4c p.174
Branch circuits, over 50 amp 210-23d page 62
Branch circuits, signs 600-5a p.690
Branch circuits, small appl. 20a 210-52b1 p.63
Branch circuits, small appliance 220-4b p. 71
Branch-circuit selection 440-4c FPN p.460
Brazed or welded 410-15b1 page 362
Breaker, height from floor 380-8a p.322
Breaker, highest location 380-8a page 322
Breakers, as switches 380-11 p.323
Breakers, as switches SWD 240-83d p.119
Breakers, maximum number 384-15 page 328
Breakers, next higher size 240-3b p.108
Breakers, standard sizes 240-6 p.110
Breakfast room receptacles 210-52b1 p.63
Bridge expansion joints 424-44c page 395
Bridge frame, grounding 610-61 p.706
Bridge wire conductors 610-21d p.703
Brittle 362-15 FPN p.293
Brittle, nonmetallic conduits 347-2 FPN p.267
Broken, parts 110-12c p.40
Building component definition 545-3 p.626
Building exceeds 3 stories 50' 225-19e p.88
Building system definition 545-3 p.626
Buildings, unfinished accessory 210-8a2 p.57
Bulk storage plants 515 p.550
Bull-switches, motion pictures 530-15d p.618
Bundled cables 520-2DEF p.600
Bundled conductors 3 or more Note 8 p.196
Burglar alarm systems 725-1 p.819
Burial depths Class I Div.I 514-8 ex.2 p.550
Burial depths minimum cover T.300-5 p.163
Buried cable over 600v depth T.710-4b p.806
Burrs and fins, fixtures 410-30a page 366
Bus bar ampacity 374-6 page 319
Bus bar spacing motors T.430-97 p.446
Bus drop cable 364-8b p.297
Busbar arrangements 384-3f page 326
Busbar, clearance insulated 384-10 page 327
Busbar, clearance noninsulated 384-10 p. 327
Busbars, insulated spacing 384-10 page 327
Busbars, motor control center 430-97 p.445
Busbars, noninsulated spacing 384-10 p. 327
Bushed hole, raceway or cable 300-16a p.169
Bushed openings 230-54e p.98
Bushing, #4 and larger conductor 300-4f p.162
Bushing, abrasion 347-12 p.269
Bushing, cable to conduit 300-5h p. 165
Bushing, in lieu of box 300-16b page 169
Busway, barriers & seals 364-25 page 299
Busway, branch circuit 364-13 p.298
Busway, disconnecting links 364-29 page 300
Busway, plug-in device 364-8b2 p.297
Busway, reduction in size 364-11 page 298
Busway, temperature rise 364-23 page 299
Busways, support 364-5 page 297
Busways, unbroken lengths 364-6 page 297
Butadiene 500-3a2 p.496
Butane 500-3a4 p.496
Bypass isolation switches 700-6 p.790
Bypassing surge current 280-2 page 156

-C-

Cabinets, air space 251 volts 373-11a3 p.317
Cabinets, damp-wet location 373-2a p.313
Cabinets, plugs or plates 373-4 page 314
Cabinets, side gutters 373-11d ex.p.317
Cabinets, side-wiring spaces 373-11d page 317
Cabinets, side-wiring spaces 373-11d ex p.317
Cabinets, wire bending space T.373-6a p. 314
Cabinets-cutout box, live parts 373-11a3 p.317
Cabinets-cutout box, weatherproof 373-2a p.313
Cabinets-cutout boxes, doors 373-11a2 p. 317
Cabinets-cutout boxes, strength 373-10b p.316
Cabinets-cutout boxes, thickness 373-10b p.316
Cabinets-cutout boxes, wet location 373-2a p.313
Cable armors 300-12 p.167
Cable assemblies 250-33 ex.2 page 131
Cable assemblies 300-12 ex. p.167
Cable bends 338-6 p.252
Cable clamps, deduction of wire 370-16b2 p.304
Cable limiters 230-82 ex.1 p.101
Cable markings, CATV T.820-50 p.875
Cable markings, Class I,II,III T.725-71 p.832
Cable markings, communication T.800-50 p.859
Cable markings, fire cable T.760-31g p.839
Cable markings, fire cable T.760-71h p.844
Cable markings, optical fiber T.770-50 p.847
Cable sheathing, protect from corr. 300-6 p.165
Cable sheaths 300-12 p.167
Cable substitutions Class I.II,III T.725-61 p.829
Cable substitutions, CATV T.820-53 p.878
Cable substitutions, comm. T.800-53 p.864
Cable substitutions, fire T.760-61d p.843
Cable substitutions, optical T.770-53 p.850
Cable tray system definition 318-2 p.216
Cable tray, single conductor #1/0 318-3b1 p.217
Cable trays, cable Class I,II,III 725-3d p.820
Cable trays, cable splices 318-8a page 219
Cable trays, covered 6 feet 318-11a ex.1 p.223
Cable trays, direction & elevation 318-5e p.218
Cable trays, edges, burrs, project. 318-5b p.218
Cable trays, elect. continuity 318-6a page 218
Cable trays, excessive movement 318-8d p.220
Cable trays, hoistways 318-4 page 218
Cable trays, maintained space 318-11b4 p.224
Cable trays, side rails 318-5d page 218
Cable trays, strength & rigidity 318-5a p.218
Cable, bus drop 364-8b p.297
Cable, deflecting junction box 300-19b3 p. 170
Cable, electric vehicle T.400-4 p.342
Cable, elevator type & uses T.400-4 page 333
Cable, irrigation 675-4 p.751
Cable, plenum T.725-71 p.829
Cable, portable power T.400-4 page 334
Cable, power-limited tray T.725-61 p.829
Cable, range and dryer T.400-4 page 340
Cable, riser Class I,II,III T.725-61 p.829
Cable, trailing 90-2b2 p.21
Cable, travel elevator note 5 T.400-4 p.332
Cablebus, conductor supports 365-3d p. 301
Cablebus, exposed work 365-2 p.300
Cablebus, through dry floors 365-6c p. 301
Cables, ampacity T.400-5a page 333
Cables, coaxial fire signaling 760-71g p.844

KEY WORD INDEX

Cables, hazardous location 725-61d p.828
Cables, heating 424-34 page 392
Cables, laid in notches of studs 300-4a2 p.161
Cables, nonheating leads 424-34 page 392
Cables, portable over 600v 400-31 p.349
Cables, stage O.C.P. 400% 530-18a p.619
Cables, traveling elevator 620-12a1 p.710
Cables, types W & G ampacity T.400-5b p.346
Cables, wet locations 310-8 page 179
Cadmium approved corrosion rest. 300-6a p.165
Calculated number of cond. .8 Tbl.1 note 7 p.879
Calculations, examples p.897-913
Calculations, mobile homes 550-13 p.641
Calculations, park trailer 552-47 p.677
Calculations, rec. veh. park T.551-73 p.665
Camping trailer definition 551-2 p.646
Can be started, motors 430-7b (1) page 418
Candelabra sockets 410-51 ex. page 369
Candelabra-base lampholders 410-27b p.365
Canned pumps 501-5f 3 p.510
Canneries, indoor wet location 300-6c p.166
Canopies, metal over 8 pounds 410-38b p.368
Canopies, space fixtures 410-10 page 361
Canopy 370-25 page 309
Canopy switches, fixtures 410-38c page 368
Capacitors 460 p.486
Capacitors, ampacity conductors 460-8a p.487
Capacitors, case not grounded 250-42 ex.3 p.132
Capacitors, connection to term. 460-28b p.489
Capacitors, discharge circuit 460-28b p.489
Capacitors, disconnection 460-28b p.489
Capacitors, dust Class II Div.I 502-2a3 p.518
Capacitors, group-oper. switches 460-24a p.488
Capacitors, induction heating 665-24 p.740
Capacitors, O.C.P. 460-25c p.488
Capacitors, overcurrent 460-8b2 p.487
Capacitors, PF correction motors 430-2 FPN p.414
Capacitors, residual voltage 460-6a page 487
Capacitors, surge-protective 502-17 p.526
Capacitors, vaults 460-2a p.486
Capitors, phase converter 455-23 p.486
Car house, railway conductors 110-19ex. p. 44
Car light, disconnect elevators 620-53 p.718
Car lights, elevator 620-22a p.714
Carbon black 500-3b2 p.497
Carbon dioxide, transformers 450-43 ex. p.482
Carnivals Article 525 p.613
Carpet squares, FCC cable 328-1 page 234
Carries the electric signal 430-71 p.438
Cartridge fuses 240-60 page 117
Catalog number, heating element 426-25 p.406
Catheter 517-11 FPN p.570
Cathode, heater or filament 640-10a p.732
CATV cable 820-51c p.876
CATV systems 820-1 p.871
CATV, coaxial grounding cond. 820-40a3 p.874
CATV, lightning conductors 820-10f3 p.872
CATVP cable 820-51a p.876
CATVR cable 820-51b p.876
CATVX cable 820-51d p.876
Cause discomfort 702-2 FPN p.800
Caustic alkalis, MC cable 334-4 p.245
Caution ... volts, tubing marking 410-91 p. 377
Caution signs, de-icing equipment 426-13 p.404

Ceiling fans, bathtubs 410-4d p. 359
Ceiling fans, support 422-18 page 383
Ceiling framing member 410-16c page 363
Ceiling support system 300-11a2 ex p.167
Ceiling support wires 300-11a p.167
Ceiling surface, heating cables 424-41e p.394
Cell explosion, batteries 480-9b p.492
Cell line conductors 668-12 p.746
Cell line definition 668-2 p.744
Cell line working zone 668-10 p.745
Cells in jars, batteries 480-5b p.491
Cellular concrete floor race., cond. 358-10 p.290
Cellular concrete floor raceway 358-2 p. 289
Cellular metal floor raceway 356-1 page 288
Cellulose fiberboard 410-76b page 375
Cellulose nitrate film vaults 530-51 p.621
Celsius of the gas or vapor 501-8b p.513
Cements, high-heat type 410-72 page 374
Center conductor T.400-4 note 6 page 344
Center contact, switched lamp 410-52 p.369
Centerline, hallway receptacle 210-52h p.65
Centralized distribution, sound 640-1 p.730
Ceramic floor, FCC cable 328-4b page 235
Chains, fixture 410-28e page 365
Chair lifts 620-1 p.701
Channel size Table 352-45 p.284
Characteristic elect. identification 780-3a p.851
Charcoal 500-3b2 p.497
Charging electrodes, spray appl. 516-6d p.565
Chemical de-icers 300-6c FPN page 166
Chimes 620-2 p.708
Chimney or flue, transformers 450-25 p.480
Chimneys 225-19b p.88
Choice of materials 310-4 FPN p.177
Christmas tree lighting 410-27b ex. page 365
Church chapels, place of assembly 518-2 p.599
Cinder fill, IMC 345-3c page 261
Cinder fill, rigid metal conduit 346-3   p. 264
Circuit breaker voltage rating 240-85 FPN p.119
Circuit breaker marked 1ø - 3ø 240-85 p.119
Circuit breaker, 4-pole 445-4e p.470
Circuit breaker, adjustments 240-82 p.118
Circuit breaker, double-coil 445-4e p.470
Circuit breaker, highest location 380-8a p. 322
Circuit breakers, as switches 380-11 page 323
Circuit breakers, as switches SWD 240-83d p.119
Circuit breakers, handles/levers 240-41b p.116
Circuit breakers, indicating 240-81 page 118
Circuit breakers, interrupt. rating 240-83c p. 118
Circuit breakers, max. number 384-15 p. 328
Circuit breakers, oil 710-21a3 p.808
Circuit breakers, standard sizes 240-6 page 110
Circuit breakers, trip free 240-80 p.118
Circuit directory, panelboard 110-22 p.44
Circuit transfer, port. switchboard 520-50b p.606
Circuits derived from transformers 215-11 p. 68
Circuits less than 50 volts 720 p.819
Circuits less than 50v, conductors 720-4 p.819
Circuits less than 50v, lampholders 720-5 p.819
Circular mil area Table 8 page 888
Circular mil area, compensate VD 250-95 p.150
Circuses Article 525 p. 613
Clamp box fill 370-16b2 p.304
Class 2 & 3 cir., wiring meth. 725-51 p.825

KEY WORD INDEX   pg - 7

Class 2 & 3 circuits 725-41 p.824
Class 20 or 30 overload relay 430-34 FPN p.431
Class I cir grd electrode cond. 250-26 ex.1 p.130
Class I circuits power limited 725-21 p.821
Class I circuits, conductors 725-27 p.823
Class I Div. I - 2 under. wiring 514-8 ex.2 p.550
Class I location sealing compound 501-5 p.506
Class I, Class 2, & Class 3 circuits 725 p.819
Class I, Div. 2 isolating switch 501-6b2 p.511
Class I, Div. 2 overcurrent prot. 501-6b4 page 511
Class I, Div. I & 2 bonding 501-16a p.516
Class I, Div. I-2 seal thickness 501-5c3 p.508
Class I, Div.2 sealing fitting 501-5b2 p.507
Class I, Div.2 wiring methods 501-4b p.505
Class I, Div.I motors & generators 501-8a p.512
Class I, Div.I wiring methods 501-4a p.505
Class I, Div.I location flr. - ceiling 517-60a2 p.590
Class I, Div.I pendant fixtures 501-9a3 p.513
Class I, Div.I seals adjacent to box 501-5a2 p.506
Class I, Group C type receptacle 517-61a5 p.591
Class I, Zone 0, 1 and 2 locations 505 p.535
Class II control circuits raceway 725-26 p.823
Class II location combustible dust 500-6 p.501
Class II locations 502 p.517
Class II max. O.C.P. 725-51 p.825
Class II, Div. I flexible connect. 502-4a2 p.519
Class I, Division I location 500-6a p.501
Class III locations 503 p.526
Class III, Div.I & 2 motors & gen. 503-6 p.527
Class III, Division 2 location 500-7b p.502
Classroom, school cooking Note 5 p.76
Clean air purging 501-8b FPN2 p.513
Clean surfaces, continuity 250-118 p.153
Cleaners 110-12c p.39
Cleaning & lub. compounds 110-11 FPN2 p.39
Clearance of bare live parts 374-7 page 319
Clearances, service drop 230-24b p.92
Clearances, signs 225-19b p.88
Clearances, working Table 110-16a page 42
Clearly and durably identified 10' 346-15c p.267
Cleat-type lampholders 410-3 ex. page 358
Cleat-type receptacles 410-3 ex. page 358
Climbing space, poles outdoors 225-14d p.86
Climbing space, through cond. 800-10a3 p.853
Clips identified for use 370-23c page 307
Clips identified for use 410-16c page 363
Clock motors 430-32c FPN p.430
Clock motors, can not damage 430-81b p.441
Clock, electric outlet 210-52b2 ex.1 page 64
Closed construction definition 545-3 p.626
Closed-loop & programmed power 780-3a p.851
Closed-loop power systems 240-20c page 111
Closet, light fixture height 410-8 p.359
Clothes closets, fixtures 410-8a page 359
Clothes closets, overcurrent device 240-24d p.115
Clothes dryer, grounded cond. 250-60 p.138
Clothes dryer, neutral 70% 220-22 p.77
Clothes dryers, demand factor T.220-18 p. 74
Clothes dryers, feeder load 220-18 page 74
Clothes hanging rod 410-8a page 359
Clothes washer/dryer Example 2b p.900
Clothes washers, rec. vehicle 551-2 FPN p.646
CM cable, communications cable 800-51d p.861
CMA, circular mil area Table 8 p.888

CMP cable 800-51a p.860
CMR cable 800-51b p.860
CMUC cable 800-51f p.861
CMX cable 800-51e p.861
CO/ALR receptacles 410-56b page 370
CO/ALR snap switches 380-14c p.324
Coal 500-3b2 p.497
Coat of plaster, heating cables 424-41c p.394
Coating processes 516 p.556
Coaxial cable grounding cond., 820-40a3 p.874
Coaxial cable separation lightning 820-10f3 p.872
Coaxial cables, fire minimum size 760-71g p.844
Code arrangement 90-3 page 22
Code enforcement, inspector 90-4 page 22
Cofficient of expansion 300-7b FPN p.166
Coil in a tank or chamber 665-44 ex.2 p.742
Coke dust 500-3b2 p.497
Cold-storage warehouses, fixtures 410-4a p. 358
Collect deposits in ducts or hoods 410-4c3 p. 359
Collector rings, definition 675-2 p.751
Collector rings, open motors 430-14b p.423
Collector rings, revolving machines 710-44 p.816
Color braid of flexible cords 400-22a p.348
Color isolated conductors 517-160a5 p.597
Color, blue 504-80c p.534
Color, blue 208v heat cable 424-35 p.393
Color, brown 277v heat cable 424-35 p.393
Color, brown isolated conductor 517-160a5 p.597
Color, gray 200-7 p.50
Color, gray 210-5 p.54
Color, gray 310-12a p.181
Color, gray 400-22a&c p.348-349
Color, green 250-119 p.153
Color, green 250-57b p.136
Color, green 250-59b p.138
Color, green 310-12b p.181
Color, green 400-23b p.349
Color, green 550-5b p. 633
Color, green 550-11 p.639
Color, light blue 504-80c p.534
Color, light blue, cords 400-22c p.349
Color, orange 215-8 p.68
Color, orange 230-56 p.98
Color, orange 384-3e p.326
Color, orange heat cable 424-35 p.393
Color, orange isolated conductor 517-160a5 p.597
Color, red 240v heat cable 424-35 p.393
Color, white 200-7 p.50
Color, white 210-5 p.54
Color, white 310-12a p.181
Color, white 400-22a&c p.348-349
Color, yellow 120v heat cable 424-35 p.393
Colored braid, flexible cords 400-22a p.348
Colored stripe 210-5 page 54
Colors, branch circuits 210-5 p.54
Colors, conductors 310-12 p.181
Colors, heating cables 424-35 page 393
Column type panelboards 300-3b ex.2 p.160
Combination electrical systems 552-20 p.671
Combustible dust Class II location 500-6 p.501
Combustible fibers Class II Div.II 500-7b p.502
Combustible material, fixture 410-65a p. 372
Combustible shades, fixtures 410-34 page 367
Combustible walls front edge of box 370-20 p.306

KEY WORD INDEX

Commercial cooking equipment T.220-20 p.76
Commercial cooking hood 410-4c1 page 359
Commercial garages Class I Div.II 511-3a p.539
Commercial garages GFCI 511-10 p.541
Commercial garages pit air changes 511-3b p.539
Common batteries 517-64b4 p. 594
Common bonding grid, cond. size 680-22b p. 764
Common grounding electrode 250-54 p.136
Common neutral 225-7b p.84
Common-return conductors, 650-5c p.736
Communication circuits, lightning 800-13 p.855
Communication cond., spacing 225-14d p. 86
Communication wires 800-51h p.861
Communication, equipment 90-2b4 p.22
Communications circuits 800 p.852
Communications, bonding elect. 800-40d p.858
Communications, electrode 800-40b1 p.858
Communiction cond. pole spacing 800-10a1 p.853
Communictions, grd. conductor 800-40a3 p.857
Community antenna grd. cond. 820-40a3 p.874
Community antenna TV and radio 820 p.871
Commutators open motors 430-14b p.423
Compact aluminum conductor Table 5a p.887
Compacted strands 501-5 FPN2 p.506
Compactor, cord length 422-8d2a page 381
Compatible with the charger 700-12a p.793
Compensate for voltage drop 250-95 p.150
Completed seal, thickness 501-5c3 p.508
Composition bushing 370-25c page 309
Compound, sealing thickness 501-5c3 p.508
Computed floor area lighting load 220-3b p.69
Computer room data process 645-2 FPN p.733
Computer room, fire cable 645-5d5 p.734
Computer systems neutral 210-4a FPN p.53
Computer/data processing 645 p.732
Concentric knockouts, bonding 250-72d p.140
Concentricity T.400-4 note 5 page 332
Concrete embedded elements 547-8b p.630
Concrete or wood above raceway 354-3a p.286
Concrete, brick or tile grounded 110-16a p.41
Concrete-encased electrode 250-81c p.145
Concretetight type fittings 345-9a p.262
Conduct safely any ground fault 250-51 p.135
Conductive materials ground 250-1 FPN2 p.120
Conductive reinforcing wire 668-31 p.748
Conductive surfaces 424-41h page 394
Conductive, optical fiber cables 770-4b p.845
Conductor insul., stage lights 520-42 p.604
Conductor insulation, batteries 640-9b p.731
Conductor minimum size, fixture 410-24b p. 364
Conductor to be grounded AC 250-25 page 129
Conductor, bending radius hi-volt 300-34 p.173
Conductor, support spacing T.300-19a p. 170
Conductor, supports 320-6 p.226
Conductors #8 & larger stranded 310-3 p.176
Conductors are laid in place 362-1 page 291
Conductors entering a building 300-5d p.162
Conductors in raceways 300-17 page 169
Conductors of the same circuit 300-3b p.159
Conductors, 60° C 110-14c1 p. 41
Conductors, 75° C 110-14c2 p. 41
Conductors, AC-DC same conduit 300-3c1 p.160
Conductors, adjacent load-carry 310-10(4) p.179
Conductors, aluminum size 310-14 p.182

Conductors, antenna clearance 810-54 p.869
Conductors, arc projector 540-13 p.625
Conductors, bare Note 5 page 195
Conductors, braided-covered 710-5 p.807
Conductors, bridge wire 610-21d p.703
Conductors, bundled Note 8 p.196
Conductors, buried minimum cover 300-5 p.162
Conductors, cell line 668-12 p.746
Conductors, choice of material 310-4 FPN p.177
Conductors, Class 1 circuit 725-2 p.820
Conductors, common return 650-5c p.736
Conductors, considered outside 230-6 p.91
Conductors, contact installation 610-21 p.702
Conductors, controllers, resistors 430-32d p.430
Conductors, copper-clad Alum. T.310-16 p.191
Conductors, corrosive conditions 310-9 p. 179
Conductors, different insulations 310-10 p.179
Conductors, direct burial 310-7 page 178
Conductors, dissimilar metals 110-14 p.40
Conductors, electroplating 669-5 p.748
Conductors, emerging from ground 300-5d p.162
Conductors, fire circuits 760-27 p.836
Conductors, free air T.310-17 p.192
Conductors, grounding size T. 250-95 page 151
Conductors, grouped together 300-20a p. 171
Conductors, heat dissipates 310-10(3) p. 179
Conductors, heat generated 310-10(2) p. 179
Conductors, identification 310-12 page 181
Conductors, in conduit Tables C1-12A p.941-1018
Conductors, in parallel 310-4 page 176
Conductors, lead covered 310-8a page 179
Conductors, less than 50v 720-4 p.819
Conductors, low & high voltage 300-32 p. 173
Conductors, marker tape 310-11b2 page 180
Conductors, marking requirement 310-11 p.180
Conductors, methods of const. 310-4 FPN p.177
Conductors, minimum size T.310-12 page 178
Conductors, nonshielded 310-6 ex. page 178
Conductors, number and size 300-17 p.169
Conductors, oil resistant T. 310-13 page 183
Conductors, on poles supports 225-14d p.86
Conductors, open service T.230-51c p.97
Conductors, operating temperature 310-10 p.179
Conductors, organs 650-5a p.736
Conductors, orientation of 310-4 FPN p. 177
Conductors, outside a building 230-6 p.91
Conductors, overhead pool 680-8 p.758
Conductors, ozone resistant 310-6 page 178
Conductors, parallel #2 310-4 ex. 4 p.177
Conductors, pilot lights 520-53f ex. p.608
Conductors, power-limit. fire 760-27 p.836
Conductors, racked 110-12b p.39
Conductors, same circuit 300-5i p. 165
Conductors, service entrance size 230-42 p.95
Conductors, shielding 310-6 page 178
Conductors, solid dielectric 310-6 page 178
Conductors, stranded #8 in raceway 310-3 p.176
Conductors, stranded on fixt.chain 410-28e p.365
Conductors, suffixes marking 310-11c p. 181
Conductors, supp. vert. raceways 300-19 p.170
Conductors, support methods 300-19b p. 170
Conductors, surface marking 310-11b1 p. 180
Conductors, tag marking 310-11b3 page 180
Conductors, temp. limitations 310-10 p. 179

KEY WORD INDEX   pg - 9

Conductors, to be grounded 250-25 p.129
Conductors, wet locations 310-8 page 179
Conductors, within 3" of ballast 410-31 p.367
Conductors, X-ray equipment 660-6 p.737
Conductors, Xenon projectors 540-13 p.625
Conduit bodies, accessible 370-29 page 310
Conduit bodies, cross-section area 370-16c p.305
Conduit bodies, seals Class I Div.I 501-5a1 p.506
Conduit nipple, fill 60% Chapter 9 note 3 p.879
Conduit nipple, no derating Note 8 ex.3 p.196
Conduit stems Class I Div.I 501-9a3 p.513
Conduit, bend radius T.346-10 p.265
Conduit, burial depth T.300-5 p.163
Conduit, conductor fill Tables C1-12A p.941-1018
Conduit, cutting die 500-2 p.493
Conduit, fill percent Table 1 p.879
Conduit, fill percent Table 4 p.880
Conduit, IMC, taper per foot 345-8 page 261
Conduit, rigid, taper per foot 346-7b page 264
Conduit, seals Class I, Div.2 501-5b2 p.507
Conduit, seals Class I, Div.I 501-5a1 p.506
Conduits, two or more in box 370-23d   p.307
Connected conductors 110-14c FPN p.41
Connecting cables, data 645-5b p.733
Connection ahead of disconnect 700-12e p.794
Connection to terminals 230-81 page 101
Connector strip 520-2 DEF p. 600
Connector strips, stage 520-46 p.606
Considered as alive 410-80c page 376
Considered as energized 410-73b page 374
Construction sites GFCI 305-6 p.175
Contact conductors installation 610-21 p.702
Contact with bodies, volts 517-64a1 p.593
Contaminants, motors 430-13 p.423
Contaminated by foreign materials 110-12c p.40
Contiguous with hospitals 517-40c p.583
Contiguous, underfloor raceways 354-3d p.286
Continuity of the circuitry 410-105b page 378
Continuity, bonding concentric KO 250-72d p.140
Continuity, clean surfaces 250-118 p.153
Continuity, device removal 300-13b page 168
Continuity, equip. grd. conductor 250-114 p.152
Continuity, mechanical 300-12 page 167
Continuous amps exceed 3 times 240-100 p.119
Continuous between cabinets 300-12 p.167
Continuous duty, escalator 620-61b2 p.719
Continuous duty, motors T.430-22a ex. p.425
Continuous industrial process 230-95 ex.1 p.104
Continuous load, branch circuit 210-22c p.61
Continuous load, branch circuit 220-3a p. 69
Continuous load, feeders 220-10b page 72
Continuous load, panelboard 384-16c page 329
Continuous-duty F.L.C. rating 440-41b p.466
Control & signal elev. cable note 5 T.400-4 p.332
Control circuit devices torque 430-9c p.420
Control circuit devices, copper cond 430-9b p.420
Control circuit overcurrent protect. 725-23 p.822
Control drawing 504-10a FPN p.532
Control drawing, intrinsically safe 504-10a p.532
Control grid, electronic tube 640-10c p.732
Control instruments, connect. 501-3b6 p.504
Control relative humidity 424-38c page 393
Control transformer, motors 430-74b p.441
Controlled vented power fuse DEF. 100 p.36

Controller, Design E motor 430-83a ex. 1 p.442
Converter output, induction heat 665-64a p.743
Converter windings DEF 100 p.33
Converter, rec. vehicle 551-2 p.646
Conveyors or hangers, support 516-4d p.563
Cooking equip., neutral B.C. 210-19b ex 2 p. 59
Cooking units, cord connected 422-17a p. 383
Cooktops, feeder demand T.220-19 page 75
Cooling, of equipment 110-13b page 40
Coordinated O.C.P. 110-10 p.39
Coordinated thermal overload 450-3a2b p.473
Coordinated, temperature limit. 110-14c p.41
Coordination 310-15a2 p.190
Coordination, electrical system 240-12 FPN p.111
Copper bars, ampacity 374-6 p.319
Copper conductors 110-5 page 38
Copper conductors minimum size T.310-5 p.178
Copper conductors, control circuit 430-9b p.420
Copper protector grd. conductor 800-40a2 p.857
Copper-clad aluminum T.310-16 p.191
Cord & plug connect. welder 210-21b ex.2 p.60
Cord bushings, fixtures 410-44 p.368
Cord connector, cord pendant 210-50a p.63
Cord connectors and receptacles 210-7 p.55
Cord length, air conditioner 440-64 p.469
Cord length, compactors 422-8d2a page 381
Cord length, dishwasher 422-8d2a page 381
Cord length, fountain 680-51e p.771
Cord length, pool 680-7 p.758
Cord length, waste disposers 422-8d1a p. 381
Cord pendants 370-25c p.309
Cord pendants, receptacle outlet 210-50a p.63
Cord polarity, appliances 422-23 page 384
Cord, strain relief, mobile homes 550-5b p.633
Cord-connected showcases 410-29 page 365
Cords, all elastomer T.400-4 page 336
Cords, all plastic parallel T.400-4 page 339
Cords, ampacity T.400-5a page 345
Cords, appliances 422-8c p.380
Cords, attachment plugs 400-7b page 347
Cords, colored braid 400-22a page 348
Cords, colored insulation 400-22c page 349
Cords, colored separator 400-22d page 349
Cords, cooking units 422-17a page 383
Cords, ease in servicing 422-17a page 383
Cords, electric vehicle T. 400-4 p.342
Cords, extra-hard usage 410-30b page 366
Cords, extra-hard usage 501-14b p.505
Cords, extreme flexibility T.400-4 note 2 p.332
Cords, fountain length 680-51e p.771
Cords, hard service T.400-4 page 336
Cords, hard usage for fixtures 410-30b p.366
Cords, heater T.400-4 page 335
Cords, in show windows 400-11 page 348
Cords, jacketed tinsel T.400-4 page 341
Cords, junior hard service T.400-4 page 337
Cords, knotting 400-10 FPN p.347
Cords, labels 400-20 p.348
Cords, lamp T.400-4 page 333
Cords, length pool 680-7 p.758
Cords, marking 400-6 page 346
Cords, minimum size 400-12 page 348
Cords, overcurrent protection 240-4 ex.1 p.109
Cords, overcurrent protection 400-13 p.348

KEY WORD INDEX

Cords, parallel heater T.400-4 page 335
Cords, parallel tinsel T.400-4 page 341
Cords, pendants 400-7a page 347
Cords, plugging boxes 530-18d p.619
Cords, protection from damage 400-14 p. 348
Cords, room air conditioner 440-64 p.469
Cords, shall not be concealed 400-8 p.347
Cords, showcase 410-29 p.365
Cords, sign 600-10c1 p.692
Cords, size sound recording 640-7 p.731
Cords, splices not permitted 400-9 page 347
Cords, strain or physical damage 410-30b p.366
Cords, surface marking 400-22f page 349
Cords, tension 400-10 p.347
Cords, theater stages T.400-4 note 4 page 332
Cords, tinned conductors 400-22e page 349
Cords, TPT, TS, TST types T.400-4 note 2 p.332
Cords, tracer in braid 400-22b page 348
Cords, twisted portable T.400-4 page 335
Cords, uses not permitted 400-8 page 347
Cords, uses permitted 400-7 page 347
Cords, vacuum cleaner T.400-4 page 341
Cords, winding with tape 400-10 FPN p.347
Corner-grounded delta 240-83 FPN p.119
Corner-grounded delta 250-5 FPN p.123
Corrosion, deteriorating agent 110-11 p.39
Corrosion-resistant NMC cable 336-30a2 p. 250
Corrosion-resistant UF cable 339-1a page 253
Corrosive atmosphere 547-5b p.629
Counter top receptacles 210-52c p. 64
Counter top surfaces 210-8a6 p. 57
Counter-mount. cook. recpts. 210-52b2 ex.2 p.64
Counters, bar type recpts. required 210-52a p.63
Cove lighting, adequate space 410-9 p.361
Cover, boxes 370-25 p.309
Cover, buried cable definition T.300-5 p.163
Cover, faceplate or canopy 370-25 page 309
Cover, gutter 374-4 p.318
Cover, hi-voltage in concrete T.710-4b ex.1 p.806
Crane rail as conductor 610-21f4 p.703
Crane-runway track 610-21f4 p.703
Cranes Class III location 520-7a page 125
Cranes & hoists 610 p.698
Cranes or hoists, demand fact. T.610-14e p.702
Crawl space equipment 210-63 p.65
Crawl spaces, GFCI 210-8a4 page 57
Critical branch, definition 517-2 p.566
Critical care area receptacles 517-19b p.573
Cross-connect arrays 800-53c p.864
Cross-connect assemblies 800-4 ex.1 p.852
Cross-sectional area Table 4 100% p.880
Crossarms, communication cts. 800-10a2 p.853
Crossings, FCC 328-17 page 236
Crowfeet, fixtures 410-16d page 363
Crucibles, definition 668-2 p.744
CSA, cross-section area Table 4 100% p.880
Cupboards 210-52a page 63
Curing apparatus required outlets 516-2e p.561
Current collectors Class III 503-13c p.530
Current rating of the unit, fixture 410-35b p.367
Current sensing device 669-9 p.749
Current transformers, cases 250-122 ex.   p.154
Current-limiting O.C.P. device 240-11 page 110
Curtain machines, stage 520-48 p.606

Cut or abrade the insulation 410-28a page 365
Cutout boxes wet location 373-2a p. 313
Cutout boxes, plugs or plates 373-4 page 314
Cutouts and fuse links 710-21c p.810
Cutouts, flashers signs 600-6b p.691
Cutting die, conduit 345-8 p. 261
Cutting die, conduit 346-7b p.264
Cutting die, conduit 500-2 p.493
Cutting slots in metal 300-20b page 171
Cutting tables, lampholders 530-41 p.621
Cyclic operation, heat 424-22d ex. c page 392
Cyclopropane 500-3a4 p.496

-D-

Dairies, indoor wet locations 300-6c p.166
Damage to conductors 300-17 page 169
Dampers, transformers 450-45e p.483
Damping means, motors 430-52a FPN p.435
Damping transitory overvoltages 450-5c p.476
Dance halls, place of assembly 518-2 p.599
Danger High Voltage 230-203 page 105
Danger-High Voltage-Keep out 110-34c p.47
Danger-high voltage-keep out 410-43 p.816
Dangerous temperature 240-1 FPN page 107
Dangerous temperatures 240-100 FPN p. 119
Data errors, grounding 250-21d page 126
Data processing rooms, disconnect 645-10 p.734
Data processing, branch circuits 645-5 p.733
Day rooms, lounges, corridors 517-3 FPN p.568
DC cell line, process power supply 668-11 p.745
DC conductors for electroplating 669-9 p.749
DC generators, neutral conductors 445-5 p. 470
DC generators, protection 530-63 p.622
DC ground protector 665-44a p.742
DC plugging boxes 530-14 p.618
DC resistance, Table 8 p.888
DC system grounding 685-12 p.775
DC voltage, aux. rectifiers 665-24 p.741
De-icing equipment 426 page 403
Dead end open wiring 320-6 p.226
Dead ends nonmetallic wireway 362-25 p.294
Dead ends, wireways 362-10 page 292
Dead front, panelboard 384-18 page 329
Dead-front & dead-rear 520-24 p.602
Dead-front construction, plugs 410-56f p. 370
Decimal fraction .8 Table 1 note 4 p.879
Decorative bands, fixtures 410-19b ex. p. 364
Decorative lighting outfits 410-27b ex. p. 365
Decorative panels 424-42 page 395
Dedicated space 210-8a2 ex.2 page 57
Dedicated space for appliance 210-8a2 ex.2 p.57
Dedicated space, panelboard 384-4 page 326
Definition unfinished basement 210-8a5 p.57
Deflecting the cables 300-19b3 page 170
Deformed at normal temp. 310-13 FPN p.182
Degree of flexibility 351-23b3 ex.2 p.279
Deleterious effect 310-9 page 179
Deleteriously affected, bushing 445-8 p.471
Delta breakers 384-16e p.329
Delta high-leg, feeder 215-8 p.68
Delta high-leg, panelboard 384-3e p. 325
Delta high-leg, service 230-56 p.98
Demand factor, cooking equip. T.220-19 p.75

KEY WORD INDEX pg - 11

Demand factor, definition 100 page 27
Demand factor, dryers T.220-18 p.74
Demand factor, elevators T.620-14 p.712
Demand factor, farm T.220-40&41 p.82
Demand factor, fixed appliances 220-17 p.74
Demand factor, kitchen equip. T.220-20 p.76
Demand factor, laundry T.220-11 p.72
Demand factor, lighting T.220-11 p.72
Demand factor, mobile homes 550-22 p.643
Demand factor, optional DW unit T.220-30 p.77
Demand factor, optional DW units T.220-32 p.79
Demand factor, optional restaurant 220-36 p.81
Demand factor, optional school T.220-34 p.80
Demand factor, receptacles 220-13 p.73
Demand factor, shore power recpts. 555-5 p.687
Demand factor, small appliances T.220-11 p.72
Derangement, emergency systems 700-7a p.790
Derating factors trench cond. Note 8 ex.4 p.196
Design E controller 430-83a ex. 1 p. 442
Design E motor disconnect 430-109 ex. 1 p. 447
Design E motors 430-52c3 ex. 1 p. 435
Design, specification 90-1c p.21
Destructive corrosive conditions 334-4 p.245
Deter access 110-31 p.45
Deteriorating agent 110-11 p.39
Device or equipment box fill 370-16b4 p.305
Device removal, continuity 300-13b page 168
Devices of insulating material 336-21 p.250
Devices rated 800 amps or less 240-3b p.108
Devices rated over 800 amps 240-3c p.108
Diagnostic equipment, O.C.P. 517-73a p.595
Diagnostic testing 670-5 ex. p.751
Diagrams, feeder 215-5 p.67
Diameter rods, gas cable 325-20 page 232
Dielectric constant 517-160a6 p.597
Dielectric fluid, transformers 450-24 p.480
Dielectric heating definition 665-2 p.740
Dielectric strength test 550-12a p.641
Diesel engine as prime mover 700-12b2 p.793
Different circuits in same cable 725-54b p.827
Different circuits same cable 725-26 p.823
Different intrinsically safe circuits 504-2 p.531
Different potential 230-54e p.98
Diffraction and irradiation types 660-23b p.738
Diffusers, thermal shock 410-4c2 page 359
Dimmer systems 3-phase, 4-wire 518-5 p.600
Dimmers, autotransformer 520-25c p.603
Dimmers, night club 520-25a p.602
Dimmers, overcurrent protection 520-25a p.602
Dining facilities 518-2 p.599
Dipping processes 516 p.556
Direct burial conductors 310-7 page 178
Direct burial, raceway transitions 300-5j p.165
Direct burial "S" loops 300-5j p.165
Direct grade level access 210-52e page 65
Direction & elevations of runs 318-5e p. 218
Directional sign disconnect 600-6 ex. p.690
Directories of electrical 300-21 FPN p.171
Directory of circuits, panelboard 110-22 p.44
Directory, or plaque 225-8 p.85
Directory, or plaque 230-2b p.90
Discharge circuit, capacitors 460-28b page 489
Discharge lighting over 1000v 410-80a p. 376
Disconecting means, phase converter 455-8 p.485

Disconnect, A/C 440-11 p.461
Disconnect, amp rating, motor 430-110a p.448
Disconnect, appliances 422-21a page 383
Disconnect, computer/data 645-10 p.734
Disconnect, Design E motors 430-109 ex. 1 p. 447
Disconnect, dispensing pump 514-5 p.549
Disconnect, elevator 620-51 p.717
Disconnect, heat 424-19 p.389
Disconnect, high-voltage 230-205 p.105
Disconnect, hook sticks bus bar 364-12 p.298
Disconnect, irrigation machines 675-8 p.752
Disconnect, locked-rotor 430-110c3(1) p.449
Disconnect, marked indicate purpose 110-22 p.44
Disconnect, mobile homes 550-23e p.645
Disconnect, mobile homes 550-6a p.634
Disconnect, motors 430-101 p.448
Disconnect, outbuildings snap sw. 225-8c ex. p.85
Disconnect, outline lighting 600-6 p.690
Disconnect, range 422-22b p.384
Disconnect, room air conditioner 440-62a2 p. 468
Disconnect, self-service station 514-5b p.549
Disconnect, sign 600-6 p.690
Disconnect, snap switch 225-8c ex. p.85
Disconnect, welder AC-DC 630-13 p.726
Disconnect, welder motor-gen. 630-23 p.728
Disconnect, welder resistance 630-33 p.729
Disconnect, X-ray equipment 660-5 p.737
Disconnecting link, busways 364-29 page 300
Disconnecting means type 430-109 p.447
Disconnecting means, snap switch 225-8c ex p.85
Discontinued outlets 354-7 page 287
Discontinued outlets 356-7 page 288
Dishwasher, cord length 422-8d2a page 381
Dishwashers, rec. vehicle 551-2 FPN p.646
Disinfecting agents, flammable 517-60a2 p.590
Dispensing pump, disconnect 514-5 p.549
Display pools 680-1 FPN p.755
Disposers, cord length 422-8d1a p.381
Dissimilar loads 220-21 page 76
Dissimilar metals 348-1 page 270
Dissimilar metals, alum. & steel 346-1b p. 263
Dissimilar metals, conductors of 110-14 p.40
Dissipation of heat 300-17 page 169
Distinctive marking 210-5 ex.1 page 54
Distinctive marking, terminals 200-6ex.4 p. 49
Distributing frames 800-53c p.864
Distribution apparatus 250-42 ex.3 p.132
Distribution cutouts 710-21c p.810
Distribution system mobile home 550-21 p.643
Divided equally, track loads 410-102 page 378
Docks, Class I location 513-6a p.543
Does not carry the main power 430-71 p.438
Domed covers 370-16a page 303
Domestic oil burner 430-32c3 p.430
Donut-type current trans. 450-5a3 FPN p.476
Door swing, mobile homes 550-8d ex.3 p.637
Door vehicle, attached garage 210-70a p.66
Double canopy rings 410-38c page 368
Double ended 250-23a ex.4 page 126
Double insulated 250-45c ex. page 134
Double insulated shaver 422-23 ex. p.384
Double locknuts 501-16a p.516
Double set screws 410-38c page 368
Double-coil circuit breaker 445-4e p.470

Double-pole switched lampholder 410-48 p.369
Downspout, outdoors 225-13 p.86
DP cable 645-5d5 p.734
Draglines over 600v 710-41 p.815
Drain fittings 680-41d4 ex. p.769
Drainage of raceways 225-22 p.88
Drainage of raceways 230-53 p.97
Draining charge, capacitor 460-6 p.486
Draw-out type switchgear 710-22 p.812
Drawbridges T.430-22a ex. p.425
Dredges over 600v 710-41 p.815
Dressing room, pilot light 520-73 p.613
Dressing rooms, lights & recpts. 520-73 p.613
Drinking water coolers 440-13 p.463
Drip loops, services 230-54f page 98
Driveways, service height 230-24b p.93
Driving machine motors 620-61b2 p.720
Drop Box 520-2 DEF page 600
Drop boxes, stage 520-46 p.606
Dropped ceilings 384-4 FPN3 page 326
Dry board installations 424-41g page 394
Dry kraft paper tapes 325-21 page 233
Dry-niche fixture, definition 680-4 p.755
Dryer, clothes washer Example 2b p.900
Dryer, minimum load 220-18 p.74
Dryer, neutral 220-22 p.76
Dryer, use nameplate load 220-32c3 p.80
Dryers, household 220-18 p.74
Drying apparatus 516-2e p.561
Drywall, repairing 370-21 page 306
Dual fuel supplies emergency 700-12b3 p.793
Dual-winding transformer 426-2 FPN p. 404
Duct heaters 424-57 page 396
Duct heaters, turning vanes 424-59 FPN p. 396
Duct work, heaters 424-12b page 389
Ducts or plenums wiring methods 300-22b p.172
Ducts to transport dust 300-22a page 171
Ducts, wiring in 300-22 page 171
Dumbwaiter, DEF 100 Hoistway p.30
Dumbwaiters 620 p.706
Duplex receptacle 210-8a2 ex.2 p.57
Dust with water, excessive 547-5a p.629
Dust, aluminum 502-2a3 p.518
Dust, aluminum bronze powders 502-2a3 p.518
Dust, combustible Class II location 500-6 p.501
Dust, magnesium 502-2a3 p.518
Dust, transformer or capacitor 502-2a3 p.518
Dwelling example, p.897
Dwelling service size Note 3 page 195
Dwellings, over 1000v lighting 410-80a p. 376
Dwellings, receptacles *no limit T.220-3b p.70
Dynamic braking resistor 430-29 p.427

-E-

Earth resisivity 800-30a FPN4(3) p.855
Earth, equipment grd. conductor 250-51 p.135
Ease in servicing, cords 422-17a page 383
Easily ignitible fibers 500-7b p.502
Easily ignitible material 240-24d page 115
Eccentric knockouts 250-72d page 140
Effective grounding path 250-51 page 135
Effective interlocks 516-2e p.561
Effectively grounded, def. 800-30a1 FPN p.856

Egress lighting energized 410-73e ex.3 p.374
Egress lighting, emergency 700-16 p.794
Egress, illumination 517-42a p.586
Either side of the boundary 501-5b2 p.507
Ejector mechanisms, plugs 410-56g page 371
Electric cranes grounding 250-7a p.125
Electric discharge light. trans. 600-21a p.692
Electric discharge lighting 410-80a page376
Electric discharge tubing 600-41 p.694
Electric furnace transformer 450-26 ex.3 p.481
Electric furnaces 250-5b4 ex.1 page 123
Electric heat, disconnect 424-19 page 389
Electric heat, unit switch 424-19c page 390
Electric heating, branch circuit 424-3 p. 388
Electric signs 600 p.689
Electric supply company 230-204d page 105
Electric vehicle 625-2 DEF page 721
Electric vehicle cableT. 400-4 page 342
Electric vehicle cable 625-17 page 723
Electric vehicle charging 511-9 p.541
Electric vehicle markings 625-15 p. 722
Electric vehicle ventilation T. 625-29c p. 725
Electric-disc. light 1000v or less 410-73 p. 374
Electric-discharge lamp 410-54a page 370
Electric-discharge lighting 600-2 DEF p. 689
Electrical components, heaters 424-12b p. 389
Electrical contact 250-42c page 132
Electrical continuity, cable tray 318-6a p.218
Electrical continuity, FCC cable 328-11 p. 235
Electrical continuity, services 250-72 p.140
Electrical cranes Class III location 250-7a p.125
Electrical ducts 310-15d page 190
Electrical life support Def. 517-3 p.566
Electrical metallic tubing (EMT) 348-1 p. 270
Electrical nameplates 550-6d p.635
Electrical noise 250-74 ex.4 page 141
Electrical noise 300-10 ex.3 p.166
Electrical noise, receptacles 410-56c page 370
Electrical nonmetallic tubing 331-1 page 239
Electrical potential to bodies 517-64a1 p.593
Electrical rating, fixture 410-35b page 367
Electrical resonance, motors 430-2 FPN p. 414
Electrically cond. compound 300-6a ex. p.165
Electrically continuous, grounding 250-92b p.148
Electrically driven machines grd. 675-13 p.754
Electrically heated appl. signal 422-12 p.382
Electrically heated appliances 422-8a p. 380
Electrocuted, health care 517-11 FPN p.570
Electrocution 422-24 page 384
Electrode, communications 800-40b1 p.858
Electrode, diameter 250-83c2 p.146
Electrode, plate 2 sq.ft. 250-83d p.146
Electrode-type boilers 424-82 page 399
Electrodes, grounding 250-81 p.144
Electrodes, nonferrous thickness 250-83d p.146
Electroendosmosis 310-13 FPN page 182
Electrolysis of the sheath 710-32 p.814
Electrolyte, batteries 480-7b p.491
Electrolytic cell definition 668-2 p.744
Electromagnetic induction 668-3 FPN p.745
Electromagnetic interference 250-74 ex.4 p.141
Electromagnetic valve supply 650-5a p.736
Electromechanical controllers 600-2a ex.2 p.690
Electromechanical purposes DEF 100 p.35

KEY WORD INDEX　　　　　　　　　pg - 13

Electronic cameras 530-1 p.616
Electronic organs, sound 640-1 p.730
Electronic puposes DEF 100 p.35
Electronic tube 640-10c p.732
Electronically actuated fuse DEF 100 p.36
Electroplating 669 p.748
Electroplating, overcurrent prot. 669-9 p.749
Electrostatic air cleaner 422-7 ex. page 380
Electrostatic equipment 516-4b p.562
Electrostatic equipment 516-5d p.564
Electrostatic fluidized beds 516-6d p.565
Electrostatic shield 517-160a2 p.597
Electrostatic spraying 516-4 p.562
Electrostatically charged 516-4 p.562
Elevation of live parts T. 110-34e page 48
Elevator cable T.400-4 page 333
Elevator car 620-41 FPN p.716
Elevator car tops 620-85 p.720
Elevator traveling cables T.400-4 note 5 p.332
Elevator, car lights 620-22a p.714
Elevators 620 p.706
Elevators, conductors 620-11 p.709
Elevators, demand factors T.620-15 p.712
Elevators, door interlock wiring 620-11a p.709
Elevators, emergency power 620-91 p.720
Elevators, grounding 620-81 p.720
Elevators, lighting parallel 620-12a1 p.710
Elevators, machine room 620-71 p.719
Elevators, other bldg. loads 620-91b p.721
Elevators, regenerated power 620-91a p.721
Elevators, travel. cables note 5 to T.400-4 p.332
Elevators, traveling cables 620-12a1 p.710
Elliptical cross section Table 1 note 9 p.879
Embedded cables, splices 424-40 page 393
Embedded in concrete 349-4 page 273
Emergence from the ground 504-80b ex. p.534
Emergencies & for tests 305-3c page 174
Emergency circuits, boxes 700-9a p.791
Emergency illumination, egress lts 700-16 p.794
Emergency lighting circuits 700-15 p.794
Emergency power panel tapped 700-12e p.794
Emergency systems 700 p.789
Emergency systems, battery 700-12a p.793
Emergency systems, definition 517-3 p.566
Emergency systems, fuel supply 700-12b2 p.793
Emergency systems, generator 700-12b p.793
Emergency systems, loads on b.c. 700-12 p.793
Emergency systems, overcurrent 700-25 p.796
Emergency systems, power 10 sec. 700-12 p.792
Emergency systems, tests 700-4 p.789
Emergency systems, transfer switch 230-83 p.102
Emergency systems, unit equip. 700-12e p.794
Emergency systems, wiring 700-9a p.791
Emerging from the ground, cond. 300-5d p.162
EMT raceway, bends 348-10 page 272
EMT raceway, corrosion protected 348-1 p.270
EMT raceway, couplings-connectors 348-8 p.271
EMT raceway, maximum size 348-5b page 271
EMT raceway, minimum size 348-5a page 271
EMT raceway, supports 348-12 page 272
EMT, unbroken lengths 348-12 ex.2 page 272
Enamel 348-1 page 270
Encapsulated, pool 680-20b1 p.761
Enclosed overcurrent device 230-208b page 106

Enclosing leads of motors 350-10a1 page 275
Enclosure bonding 250-75 p.141
Enclosures with O.C.P. not as JB 373-8 p.315
Enclosures, arcing parts 110-18 p.44
Enclosures, in concrete 370-23f page 308
Enclosures, number of circuits 90-8b p.23
Enclosures, subsurface 110-12b p.39
End fittings clearance 384-10 page 327
End seal, MI cable Note 6 page 195
Enduring the vibration 545-13 p.628
Energized attachments 668-10a1 p.745
Energized automatically 424-20a3 page 391
Energized parts 300-31 page 173
Enforcement of code 90-4 p.22
Engine compartment wiring 551-10b5 p. 648
Engineering supervision 310-15b page 190
ENT 331-1 page 239
ENT, marking 331-15 page 242
ENT. pool lighting 680-25b1 ex. 2 p. 765
ENT, sizes 331-5 page 241
Entrap heat, fixture insulation 410-66b p. 373
Environmental air duct, plenums 300-22b p.174
Environmental conditions, fixture 410-24a p.364
Equal to or greater than 280-4 page 157
Equal to or greater than 440-2 p.458
Equalizer leads generators 445-4e p.470
Equip. grd. conductor size T.250-95 p.151
Equip. grd. conductor, earth 250-51 page 135
Equip. grd. conductor fill 370-16b5 page 305
Equip. ground fault time delay 230-95a p.104
Equipment likely energized grd. 250-42a p.132
Equipment markings 110-14c FPN p.41
Equipment rooms & pits, pools 680-11 p.760
Equipment termination prov. 110-14c1 ex. p.41
Equipment, communication 90-2b4 p.22
Equipotential bond. jumpers 501-8b FPN2 p.531
Equipotential plane 547-8b p.630
Equivalent strength boxes 370-40b ex.1 p.311
Equivalent, to splices 110-14b p.40
Equivalent, wall of equipment 110-12a p. 39
Escalator, continuous duty 620-61b2 p.719
Escalators 620 p.706
Essential b.c. panelboard bonding 517-14 p.571
Essential electrical system 517-30b1 p.575
Essential electrical system, def. 517-3 p.567
Ethanol 500-3a4 p.496
Ethylene oxide 500-3a2 p.496
Evacuation of a theater 700-12 p.792
Examination of equipment 110-3 p.38
Examples of calculations p.897-913
Excavating sidewalks 370-29 page 310
Excessive sag 342-7b1 page 257
Exciter, leads 300-3c2 ex.3 p.160
Exciter, overcurrent 445-4a p.469
Exclude moisture 501-5 p.506
Execution of work, workmanlike 800-6 p.853
Exercise of ingenuity 500-2 FPN 2 p.493
Exhaust fumes of engines 511-2 p.539
Exhaust vapors 410-4c2 page 359
Existing dwelling load additions 220-3d1 p.71
Existing installations, parallel 310-4 ex. 4 p. 177
Exit fixtures 410-73e ex.2 p.374
Exit lighting emergency systems 700-16 p.794
Exothermic welding 250-81 ex. page 144

KEY WORD INDEX

Exothermic welding 547-8a ex.2d p.630
Exothermic welding process 250-81 ex.1 p.144
Expansion characteristics PVC 362-23 FPN p.294
Expansion joints, bonding 250-77 page 142
Expansion joints, raceways 300-7b page 166
Expansion, busways 364-28 p.300
Expansion, gutters live parts 374-7 p.319
Expansion, PVC Table 10 p.890
Explanatory material FPN 90-5 p.22
Explosion proof definition 100 p.28
Exposed ballasts 410-76a page 375
Exposed busbars, service disconnect 300-16b p.169
Exposed cable wiring 336-21 page 250
Exposed conductive surfaces, def. 517-3 p.567
Exposed current to bodies 517-64a1 p.593
Exposed live parts, porcelain fixt. 410-46 p.369
Exposed non-curr. carry parts 250-42 ex.2 p.132
Expulsion type fuse links 710-21c p.810
Extended beyond the outside walls 300-5c p.162
Extension cord sets 305-6b1 p.176
Extension rings 370-16a page 303
Extension through walls 362-9 page 292
Extensions from box 370-22 p.307
Extensions from wireways 362-11 page 292
Extensions, underplaster 352-5 p. 281
Extensive metal on buildings 250-44 FPN p.133
External means for adjusting 240-6 page 110
External sensing elements 430-125c1a p.451
External spark, batteries 480-9a p.492
Extra-hard usage cords arc lamps 520-61 p.611
Extra-hard usage cords type 520-68a p.612
Extra-hard usage cords, ampacity T.520-44 p.605
Extra-hard usage, Class I 501-4b p.505
Extreme cold 347-2 FPN page 267
Extreme cold 362-15 FPN p.293
Extruded thermoplastic covering 342-1 p.255

-F-

Face up position receptacles 210-52d p.65
Face up position recept. trailers 552-41f2 p.673
Face-up receptacles 550-8f2 p.637
Faceplate thickness, switches 380-9 p.322
Faceplate, boxes 370-25 page 309
Faceplates, receptacles 410-56d page 370
Faceplates, thickness 410-56d p.370
Facilitate cleaning hood fixture 410-4c3 p.359
Facilitate overcurrent device 250-1 FPN2 p.120
Facilitate removal or disconnection 422-8c p.380
Factory assembled structure 550-2 page 631
Factory-installed internal wiring 90-7 page 23
Factory tests, park trailer 552-60 page 684
Fairs Article 525 page 613
Falling dirt, motor enclosure T.430-91 p.444
Fan circuit interlock 424-63 page 396
Fans, ceiling support 35 pounds 422-18 p.383
Fans, paddle outlet box 370-27c p.309
Far away as possible, vent trans. 450-45a p.483
Farm buildings conductors 225-19d p.88
Farm loads 220-40 page 82
Fault hazard current 517-160b p.598
Fault hazard current, definition 517-3 p.567
FC assemblies tap devices 363-10 p.295
FC cable definition 363-1 page 294

FC cable, protective covers 363-18 page 296
FC cable, taps 363-10 page 295
FC cable, terminal block ident. 363-20 p.296
FCC, cable connections 328-11 p.235
FCC, carpet squares 328-1 page 234
FCC, construction 328-30 page 237
FCC, corrosion-resistance 328-33 page 237
FCC, crossings 328-17 page 236
FCC, definition 328-2 p.234
FCC, insulating ends 328-11 p.235
FCC, marking 328-31 page 237
FCC, metal components 328-33 p.237
FCC, polarization 328-20 page 236
FCC, receptacle 328-14 p.236
FCC, release-type adhesive 328-10 p.235
FCC, shields 328-35 page 237
FCC, system height 328-18 page 236
FCC, transition assembly 328-15 p.236
FCC, voltage 300v 328-6a p.235
Feeder tap, 10 feet 240-21b page 112
Feeder tap, not over 25 feet 240-21d page 112
Feeder tap, over 25' to 100' 240-21e page 113
Feeder taps, motors 430-28 p.427
Feeder, ampacity minimum 215-2a p.67
Feeder, com. neutral same raceway 215-4b p.67
Feeder, common neutral 215-4a page 67
Feeder, demand fact. elevators T.620-14 p.712
Feeder, diagram 215-5 page 67
Feeder, GFCI equipment 215-10 page 68
Feeder, GFCI personnel 215-9 page 68
Feeder, minimum size 215-2 page 67
Feeder, mobile homes 550-24 p.645
Feeder, over 600v O.C.P. 240-100 p.119
Feeder, permanently installed 550-5a p.632
Feeder, tapped two-wire circuits 215-7 pg.68
Feeders, floating buildings 553-7b p.685
Feeders, motors 430-24 p.426
Feeders, shore power 555-5 p.687
Feeders, studios 400% 550-18b p.619
Feeders, X-ray 660-6b p.737
Female fitting, showcase 410-29c page 366
Fence 110-31 p.45
Fence fabric 110-31 p.45
Ferromagnetic envelope definition 426-2 p.404
Ferromagnetic, single conductor 427-47 p. 413
Ferrous channel raceways 352-42 page 283
Ferrous enclosures 330-16 p.239
Ferrous metal conduit, corrosion 346-1c p.263
Ferrous metal, faceplates 410-56d page 370
Ferrous raceways corrosion prot. 300-6a p.165
Festoon lighting 225-6b page 84
Festoon supports 225-13 page 86
Festoon wiring, fixtures 410-27a page 365
Festoon wiring, staggered 520-65 p.611
Festoons, stage 520-65 p.611
Fexible connect., Class II, Div.I 502-4a2 p.519
Fiberboard, cellulose low-density 410-76b p.375
Fibers Class III location 503-1 p.526
Field bends, lead sheath T.346-10 ex. p. 266
Field punched metal studs 300-4b1 page 161
Field-wiring chamber 680-25b2 p.765
Film & film scrap 530-41 p.621
Film or tape 530-1 p.616
Filter capacitors induction heat 665-24 p.741

KEY WORD INDEX   pg - 15

Final conductor span   225-19 ex.4 p.87
Fine print note (FPN)   90-5   p.22
Finished ceilings, paint, wallpaper 424-42 p. 395
Fire alarm, supply side   230-94 ex.4   page 103
Fire barriers, busway   364-25   page 299
Fire circuits, grounding   760-6   p.834
Fire circuits, conductors   760-27   p.836
Fire circuits, diff. cir. same cable 760-26 p.836
Fire circuits, overcurrent   760-23   p.835
Fire circuits, power limitations   760-41   p.839
Fire circuits, wiring methods   760-30   p.837
Fire detectors   760-55   p.842
Fire escape, conductors   225-13   p.86
Fire escapes, vent transformer   450-45a   p.483
Fire hazard analysis   450-24 FPN page 480
Fire point, transformer liquid   450-24   p.480
Fire protective signaling systems   760   p.834
Fire pump circuit   240-3a   p.108
Fire pump, supply side   230-94 ex.4   page 103
Fire pumps Article 695 page 784
Fire pumps   230-95 ex.2   p. 104
Fire pumps, disconnects   230-72b   page 100
Fire pumps, separate service   230-2 ex.1   p. 89
Fire rating, fire dampers 450-45e page 483
Fire resistance CATV cables   820-49   p.875
Fire resistance communication wire 800-49 p.859
Fire resistive ratings   300-21 FPN   p.171
Fire retardant fibrous coverimg   333-20   p. 244
Fire stop, all holes   300-21   p.171
Fire-rated floor   300-11a   p.167
Fire-resistant, low smoke   300-22c   p.172
Fire/smoke dampers   645-2(2)   p.733
Fireplaces, wall space   210-52a   page 63
First floor of building   336-5a1   page 248
First-make, last-break plug   250-99a   p.151
Fish the cables 300-4d ex.2   page 161
Fish the cables 336-4c page 248
Fittings design. for the purp.   400-10 FPN   p.347
Fittings, aluminum to steel tube   348-1 ex.   p.270
Fittings, insulated   373-6c   p.315
Fittings, support 50 pounds   410-16a   p.363
Fittings, telescoping   502-4b1   p.520
Fittings, where required   300-15c   page 169
Five threads fully engaged   501-4a   p.505
Fixed appliance, definition   550-2   p.631
Fixed appliances, feeder load 75%   220-17   p.74
Fixed wiring substitute, cords as   400-8   p.347
Fixture as raceways   410-31 ex.1   page 367
Fixture canopy, box cover   410-12   page 361
Fixture chains, stranded conduct.   410-28e p.365
Fixture exposed within hood   410-4c3   page 359
Fixture hangers, flat cable   363-12   p.295
Fixture load current   410-30c3   page 367
Fixture stems, arms or stems   410-28c   p. 365
Fixture stems, splices and taps   410-28d   p. 365
Fixture studs, boxes   370-16b3   p.305
Fixture studs, not part of box   410-16d   p.363
Fixture wire SF-1 voltage   T.402-3   p.354
Fixture wire, ampacity   402-5   p.357
Fixture wires   402   page 350
Fixture wires, marking   402-9b   page 357
Fixture wires, minimum size   402-6   page 357
Fixture wires, protection   240-4   page 109
Fixture wires, uses permitted   402-10   page 357

Fixture, adjacent comb. material   410-68 p.373
Fixture, adjustable   410-30b   page 366
Fixture, arms & stems   410-38a   page 368
Fixture, autotransformer   410-78   page 375
Fixture, ballast install fiberboard 410-76b p.375
Fixture, ballast marking   410-35a   page 367
Fixture, canopy switches   410-38c   page 368
Fixture, clothes closet   410-8d   p.361
Fixture, combustible material   410-65a   p. 372
Fixture, combustible shades   410-34   page 367
Fixture, conductor insulation   410-24a   p. 364
Fixture, conductor size   410-24b   page 364
Fixture, considered as alive   410-80c   page 376
Fixture, construction   410-68   p.373
Fixture, crowfeet   410-16d   page 363
Fixture, DC current   410-74   page 374
Fixture, design & material   410-36   page 367
Fixture, dimension   410-15a   page 362
Fixture, disconnection   410-81a   page 376
Fixture, electric-discharge   410-76a   p.375
Fixture, electrical rating   410-35b   page 367
Fixture, end-to-end assembly   410-31 ex.2   p. 367
Fixture, exposed ballasts   410-76a   page 375
Fixture, exposed metal parts   410-19b ex.   p.364
Fixture, fastened to ceiling   410-16c   p.363
Fixture, flush & recessed   410-64   page 372
Fixture, glass lamps   410-19b ex.   page 364
Fixture, handhole 410-15b1   page 362
Fixture, heavy-duty track   410-103   page 378
Fixture, hickeys   410-16d   page 363
Fixture, high-intensity discharge   410-73f   p.374
Fixture, housing construction   410-69   page 373
Fixture, insulating joints   410-16e   page 363
Fixture, lamp replacement   410-82   page 376
Fixture, lamp wattage marking   410-70   p. 374
Fixture, lens pool 18"   680-20a3   p.761
Fixture, mechanical strength   410-38   page 368
Fixture, metal canopies   410-38b   page 368
Fixture, movable or flexible parts 410-28e   p.365
Fixture, multicircuit track   410-102   page 378
Fixture, nonmetallic   410-37   page 368
Fixture, polarization   410-23   page 364
Fixture, recessed clearance   410-66a   page 373
Fixture, recessed fluorescent   410-8b2   page 360
Fixture, recessed incandescent   410-8b1   p. 360
Fixture, screw-shell max. weight 410-15a   p.362
Fixture, stranded conductors   410-28e   page 365
Fixture, supports   410-15a   page 362
Fixture, surface-mounted   410-8b1   page 360
Fixture, tap conductors   410-67c   page 373
Fixture, temperature   410-68   page 373
Fixture, tension   410-28f   page 365
Fixture, thermal insulation   410-66b   page 373
Fixture, track conductors   410-105a   page 378
Fixture, track fastening   410-104   page 378
Fixture, track load   410-102   page 378
Fixture, tree mounting   410-16h   page 363
Fixture, tripods   410-16d   page 363
Fixture, weight 50 pounds   410-16a   page 363
Fixture, weight 6 pounds   410-15a   page 362
Fixture, whip flex. conduit 250-91b ex.1   p.147
Fixture, wiring space   410-39   page 368
Fixtures, above bathtubs 410-4d page 359
Fixtures, aiming or adjusting   410-30b   p.366

KEY WORD INDEX

Fixtures, decorative bands 410-19b ex. p.364
Fixtures, in clothes closets 410-8a page 359
Fixtures, in ducts or hoods 410-4c page 359
Fixtures, in show windows 410-7 page 359
Fixtures, inspection of 410-16b page 363
Fixtures, near combust. material 410-5 p. 359
Fixtures, no-niche 680-20d p.762
Fixtures, over bathtub 410-4d p.359
Fixtures, over combust. material 410-6 p. 359
Fixtures, over vehicle lanes 511-7b p.541
Fixtures, suspended ceilings 410-16c page 363
Fixtures, temperature 90° C 410-5 p.359
Fixtures, wet locations 410-4a page 358
Fixtures, wet-niche 680-25b1 p.765
Flame arrestor, batteries 480-9a p.492
Flame-retardant covering 339-1a p.253
Flame-retardant marking 340-6 p.255
Flame-retardant NMC cable 336-30a1 p. 250
Flame-retardant, hoistway wiring 620-11a p.709
Flammable anesthetics stored 517-60a2 p.590
Flammable anesthetics, definition 517-3 p.567
Flammable disinfecting agents 517-60a2 p.590
Flammable gases or vapors 500-5a p.500
Flammable oil 710-14 p.808
Flammable paints are dried 516-2e p.561
Flanged surface inlet, fixtures 410-30c3 p.367
Flanged surface inlets, stage 520-53m p.610
Flash point, transformer 450-24 p.480
Flashers, cutouts 600-6b p.691
Flashers, switches 600-6b p.691
Flat cable assemblies, protection 363-18 p. 296
Flat conductor cable 328-1 page 234
Flat-top raceways 354-3a page 286
Flatirons, temperature limiting 422-13 p.382
Flex. cords & cables protection 400-14 p.348
Flex. metal cond., equip. grd. 250-91b ex.1b p.147
Flex. metal cond., number of cond. T.350-12 p.276
Flexible cord pendants 370-25c page 309
Flexible cords and cables 400 page 332
Flexible cords, industrial 501-11 page 515
Flexible cords, protection 240-4 page 109
Flexible metal conduit 350 page 274
Flexible metal conduit, motors 430-145b p. 453
Flexible metal conduit, supports 350-4 p.275
Flexible metallic tubing 349-1 page 273
Flexible metallic tubing, min. size 349-10a p.273
Flexural capability 426-21d page 405
Floating buildings 90-2a1 page 21
Floating buildings, wiring meth. 553-7b p.685
Floating dock, service location 555-10 p.688
Floor boxes 250-74 ex.3 p.141
Floor boxes 370-27b page 309
Floor boxes 18" to wall 210-52a p. 63
Floor covering, under heat panels 424-99b p.402
Floor or working platform, switch 380-8a p.322
Floor receptacle 210-52a page 63
Floor receptacles, protection 410-57d page 371
Floor to structural ceiling 384-4 p.326
Fluidized beds, spray area 516-6d p.565
Fluorescent fixt., as raceway 410-31 ex.1 p.367
Fluorescent fixt., thermal protect. 410-73e p.374
Fluorescent fixt., total amp rating 210-22b p.61
Fluoroscopic, X-ray 660-23a p.738
Flush lights, clearance 410-66 p.373

Flush lights, temperature limit 410-65a p.372
Flush mount recept. faceplate 410-57e p. 371
Flying material, motors 430-16 p.423
Flyings, Class III 503-1 p.526
Foot switches, induction heat 665-47b p.742
Footlight 520-2 DEF page 601
Footlights, stage 520-41 p.604
Formal interpretations of the Code 90-6 p.23
Forming shell, definition 680-4 p.756
Forming shell, pool 680-20b1 p.761
Fountain, bonding 680-53 p.772
Fountain, cord length 680-51e p.771
Fountain, GFCI 680-51a p.771
Fountain, grounding 680-54 p.772
Fountain, pumps 680-51b p.771
Fountains 680-50 p.770
FPL cable 760-71f p.844
FPLP cable 760-71d p.843
FPLR cable 760-71e p.844
Fractions of ampere, B. examples p.897
Frame of a vehicle, grounding 250-6b p.124
Frames of ranges & dryers 250-60 page 138
Frames of stationary motors 430-142 p.452
Framing member, cable support 300-4d p.161
Framing members, as support 370-23c p.307
Free air, conductors T.310-17 p.192
Free circulation of air 110-13b p.40
Free conductor at box, length 6 300-14 p.168
Free from short circuits 110-7 p. 39
Free nonheating leads 424-43a page 395
Freedom from hazard 90-1a p. 21
Freezing of piping 422-29 p.387
Freight station circuits 110-19ex. page 44
Frequencies of 360hz and higher 310-4 ex.3 p.177
Frost heaves or settlement 300-5j p.165
Fuel supply, emerg. systems 700-12b2 p.793
Fuel tanks, aircraft 513-2c p.542
Full-load current 1 ø motors   T.430-148 p.454
Full-load current 2 ø motors   T.430-149 p.455
Full-load current 3 ø motors   T.430-150 p.456
Full-load current D.C. motors T.430-147 p.453
Fumes, exposed to 110-11 page 39
Functionally associated 725-26b page 823
Fungus-resistant NMC cable 336-30a2 p. 250
Furnace transformer 450-26 p.481
Furniture, permanent 210-60ex. page 65
Fuse links, expulsion type 710-21c p.810
Fuse, controlled vented power Def 100 page 36
Fuse, Edison-base type 240-51b page 117
Fuse, electronically actuated DEF 100 p.36
Fuse, marking 240-50b page 116
Fuse, maximum voltage 240-50a page 116
Fuse, vented power definition 100 page 37
Fused switches 380-17 p.324
Fuseholders 240-60 p.117
Fuseholders, deenergized 710-21b6 p.810
Fuseless protectors 800-30a1 p.855
Fuses or breakers in parallel 240-8 page 110
Fuses that expel flame 710-21b5 p.809
Fuses, adapters 240-54 p.117
Fuses, cartridge 240-60 page 117
Fuses, disconnecting means 240-40 page 116
Fuses, in clothes closets 240-24b p.115
Fuses, next higher size 240-3b p.108

KEY WORD INDEX pg - 17

Fuses, next higher size motor 430-52a ex.1 p.434
Fuses, parallel 240-8 ex. p.110
Fuses, plug 240-50 p.116
Fuses, readily accessible 240-24a p.115
Fuses, standard sizes 240-6 page 110
Fuses, Type S 240-53a page 117
Fusible link cable connector 450-6a3 p.477
Fusion apparatus, spray area 516-2e p.561
Future expansion and convenience 90-8a p. 23

-G-

G or GR green or ground 250-119 p.153
G or GR green or ground 410-58b4 page 372
Gallery, live parts 430-132b p.452
Galvanic action, EMT 348-1 page 270
Galvanic action, IMC conduit 345-3a page 260
Galvanic action, rigid conduit 346-1b page 263
Garages snap switch disconnect 225-8c ex. p.85
Garages, above floor level 18 511-3a p.539
Garages, commercial 511 p.539
Garages, parking 511-2 p.539
Garages, pit or depression 511-3b p.539
Garages, portable light 511-3f p.540
Garages, residential GFCI 210-8a2 p.57
Garages, residential receptacles 210-52g p.65
Garbage disposal, cord length 422-8d1a p.381
Gas fired range receptacles 210-52b2 ex.2 p.64
Gas pipe, grounding 250-83a p.145
Gas pipe, in raceways 300-8 p.166
Gas pressure, IGS cable 20 pounds 325-21 p.233
Gas range, mobile home 550-11b3 p.640
Gas spacer cable, ampacity T.325-14 page 232
Gas station, double-pole switch 514-5a p.549
Gas tubing support, signs 600-41b p.694
Gases, exposed to 110-11 page 39
Gasketed busways 501-4b p.505
Gasketed wireways 501-4b p.505
Gasketed-type fixtures in ducts 300-22b p.172
Gasoline 500-3a4 p.496
Gasoline dispensing stations 514 p.545
Gasoline pump, grounded conductor 514-5 p.549
General care area receptacles 517-18b p.572
General lighting loads T.220-3b page 70
General lighting, demand factors T.220-11 p.72
General-purpose enclosures 501-6b2 p.511
Generator, ampacity 445-5 p.470
Generator, balancer sets 445-4d p.470
Generator, field control 620-13 FPN p.710
Generator, less than 65 volts 445-4c p.470
Generator, output 665-44a p.742
Generator, standby system 701-11b p.798
Generator, vehicle mounted 250-6b p.124
Generators, DC neutral conductors 445-5 p. 470
GFCI, auto diagnostic equip. 511-10 p.541
GFCI, bathrooms 210-8a1 page 57
GFCI, basements 210-8a5 p.57
GFCI, boathouses 555-3 p.686
GFCI, commercial bathrooms 210-8b1 p.57
GFCI, commercial garages 511-10 p.541
GFCI, construction sites 305-6 p.175
GFCI, counter top surfaces 210-8a6 p.57

GFCI, crawl spaces 210-8a4 p.57
GFCI, dwelling units 210-8 p.57
GFCI, fountain 680-51a p.771
GFCI, garages dwelling 210-8a2 p.57
GFCI, industrial bathrooms 210-8b1 p.57
GFCI, kitchen sink 210-8a6 p.57
GFCI, marinas 555-3 p.686
GFCI, mobile homes 550-8b p.636
GFCI, outdoor portable sign 600-10a2 p.692
GFCI, outdoors dwelling 210-8a3 p.57
GFCI, park trailers 552-41c page 672
GFCI, patient room toilet and basin 517-21 p. 574
GFCI, performance testing 230-95c page 104
GFCI, pools 680-5b p.757
GFCI, pools storable 680-31 p.767
GFCI, receptacles on roofs 210-8b2 p.57
GFCI, replacement receptacles 210-7d2 p.56
GFCI, rooftops 210-8b2 page 57
GFCI, service disconnect 230-95b FPN 2 p. 104
GFCI, service setting 230-95a page 104
GFCI, signs portable 600-10a2 page 692
GFCI, solidly grounded wye 230-95 page 104
GFCI, spa/hot tub 680-41a3 p.769
GFCI, spas and hot tubs 680-42 p.770
GFCI, storage areas 210-8a2 page 57
GFCI, unfinished accessory bldgs. 210-8a2 p. 57
GFCI, unfinished basements 210-8a5 p.57
GFCI, wet bar sink 6 feet 210-8a7 p.57
GFCI, will afford no protect. 230-95 FPN1 p.104
Glass knobs, open conductors 225-12 p.86
Glass lamps, fixtures 410-19b ex. page 364
Glue houses, corrosive prot. 300-6c FPN p.166
Gongs 620-2 p.708
Gooseneck, services 230-54b p.97
Governmental bodies 90-4 page 22
Grade level access 6' 6" 210-52e p.65
Grain drying systems T.220-41 note p.82
Gratings, vault door 450-45c p.483
Grd. elect. cond., metal enclosures 250-92b p.148
Grease, oil or cooking vapors 410-4c2 page 359
Green-colored conductor 250-59b page 138
Green-colored conductor 310-12b page 181
Green-colored hex headed screw 410-58b1 p.372
Green-colored rigid ear adapter 410-58b3 p.372
Greenfield conduit 350 page 274
Grommets, metal studs 300-4b1 page 161
Ground as grounding 250-61 page 138
Ground clamps 250-113 p.152
Ground clamps, connection 250-115 p.152
Ground clamps, protection 250-117 p.153
Ground current detection 600v 710-72d p.816
Ground detectors 250-5b ex.4 FPN page 124
Ground fault equipment time delay 230-95a p.104
Ground fault protect. service 230-95b FPN2 p.104
Ground fault protection test on site 230-95c p.104
Ground movement 300-5j p.165
Ground return, trolley wires 110-19 p.44
Ground ring depth 2 1/2' 250-81d page 145
Ground rod connection 250-112 p.152
Ground rod driven depth 250-83c3 p.146
Ground rod in soil 8' 250-83c3 p.146
Ground rod resistance 25 ohm 250-84 p.146
Ground rod size 250-83c2 p.146
Grounded cond. disc. simult. 380-2b ex.1 p.321

KEY WORD INDEX

Grounded cond., change in size 240-23 pg. 114
Grounded conductor 210-6c2 p.55
Grounded conductor, identification 200-6 p. 49
Grounded conductor, insulated 200-2 p.49
Grounded metal barrier 680-5a p.757
Grounded metal cover, heat 427-23b page 412
Grounded shield, transformer 426-32 page 407
Grounded, clothes dryer 250-60 p.138
Grounded, concrete, brick or tile 110-16a p. 41
Grounded, effectively 250-81b p.145
Grounded, effectively DEF 100 p.29
Grounded, ranges 250-60 p.138
Grounded, service cond. min. size 250-23b p.127
Grounding autotransformers, 450-5a1 page 475
Grounding cond., communication 800-40a3 p.857
Grounding conductor aluminum 250-92a p.148
Grounding conductor connections 250-113 p.152
Grounding conductor, coaxial 820-40a3 p.874
Grounding conductor, remote panel 680-25d p.766
Grounding conductor, sizing T. 250-95 page 151
Grounding connections 250-23a p.126
Grounding elect. cond. connection 250-112 p.152
Grounding elect. cond. connection 250-23a p.126
Grounding elect. cond. joint/ splice 250-91a p.146
Grounding elect. cond. protection 250-92a p.148
Grounding electrode 25 ohms 250-84 p.146
Grounding electrode bare #4 20' 250-81c p.145
Grounding electrode bonding 250-86 FPN2 p.146
Grounding electrode conductor size 250-94 p.149
Grounding electrode in soil 8' 250-83c3 p.146
Grounding electrode system 250-81 page 144
Grounding electrode, min. size 250-83c2 p. 146
Grounding electrodes mult. 250-83 p.145
Grounding of field winding DEF 100 p.31
Grounding or bond. cond. derating Note11 p.197
Grounding sheath or braid 426-22a page 405
Grounding, bridge frame 610-61 p.706
Grounding, motor control center 430-96 p.445
Grounding, pool equipment 680-24 p.765
Grounding, trolley frame 610-61 p.706
Grounding-pole identification 410-58b p. 372
Group-operated switches 600v 460-24a p.488
Grouped cables 520-2 DEF page 601
Guard strips, cable protection 333-12a p.244
Guarded DEF 100 p.30
Guards, portable handlamps 410-42b p.368
Gutters, bare live parts 374-7 page 319
Gutters, busbar ampacity 374-6 page 319
Gutters, construction 374-9b page 319
Gutters, extend beyond equip. 30' 374-2 p.318
Gutters, number of conductors 30 374-5 p.318
Gutters, splices and taps 75% 374-8a p.319
Gutters, support 374-3 page 318
Guys or braces, service mast 230-28 p.93

-H-

Habitable rooms 210-8a5 p.57
Hair dryer, cord protection 422-24 p.384
Half-round raceways 354-3a page 286
Hallway, receptacle 10' 210-52h p.65
Halon sprinklers 450-43a ex. p.482
Hand gun, sprayer 516-5b ex. p.564
Hand-guided tools, grounding 250-59 ex. p.138

Handhole for metal pole 410-15b1 p. 362
Handlamps, construction 410-42b p.368
Handle ties 240-20b ex. p.111
Handle ties, breakers 230-71b page 99
Handles or levers, breakers 240-41b p.116
Hangars, aircraft definition 513-1 p.542
Hard service cord, splices 400-9 page 347
Hard service cords T.400-4 page 336
Hard-rubber bushing 370-25c page 309
Hard-service type cord 410-29a page 366
Hard-usage, cords for fixtures 410-30b p.366
Harmonic currents 310-10 (2) page 179
Harmonic currents, busways 364-24 page 299
Harmonic currents, cords 400-5 page 345
Harmonic currents, triplen 310-4 FPN page 177
Harmonic neutral currents 210-4 FPN p.53
Harmonic neutral currents 210-4a FPN p.53
Harmonic neutral currents 220-22 FPN 2 p.77
Harmonic neutral currents Note 10c p.197
Hazard current, definition 517-3 p.567
Hazard, free from 90-1b page 21
Hazard, OCP 384-32 ex. 1 page 330
Hazard, power loss O.C.P. 240-3a p.108
Hazardous location, cables 725-61d p.828
Hazardous locations 500 p.493
Hazardous, temp. ID numbers T.500-3d p.499
Header, definition 356-1 page 288
Header, definition 358-2 page 289
Headroom, equipment 6 1/2' 110-16f p.43
Health care facilities, definition 517 p.567
Health care low volt. bodies 10v 517-64a1 p.593
Heart muscle 517-11 FPN p.570
Heat tape outlet 550-8g p.637
Heat tape outlet, park trailer 552-41d page 673
Heat, branch circuit rating 424-3 page 388
Heat, dissipation of conductors 300-17 p.169
Heated appliance signal 422-12 page 382
Heated appliances, cord 50w 422-8a p.380
Heated ceilings, wiring clearance 424-36 p.393
Heater cord T.400-4 page 335
Heater or filament, cathode O.C.P. 640-10a p.732
Heater, air flow 424-59 page 396
Heater, cords over 50 watt 422-8a p.380
Heater, disconnect 424-19 page 389
Heater, unit switch marked off 424-19c p.390
Heaters, duct work 424-12b page 389
Heaters, electrical components 424-12b p.389
Heaters, pool 680-9 p.760
Heaters, wired sections 424-12b page 389
Heating blankets 427-2 FPN page 409
Heating cable 427-2 FPN page 409
Heating cable, inspection & tests 424-45 p.395
Heating cables, concrete 16 1/2w 424-44a p.395
Heating cables, adjacent runs 424-41b p. 394
Heating cables, ceiling surface 424-41e p.394
Heating cables, construction 424-34 page 392
Heating cables, crossing joists 424-41j p. 394
Heating cables, dryboard 424-41 page 394
Heating cables, gypsum board 424-41c p. 394
Heating cables, in plaster 424-41 page 394
Heating cables, joists 424-41i page 394
Heating cables, lead wires 424-35 p.393
Heating cables, length 7' leads 424-34 page 392
Heating cables, marking 424-35 p.393

KEY WORD INDEX pg - 19

Heating cables, paint 424-42 p.395
Heating cables, secured 16 424-41f p.394
Heating cables, splices 424-41d page 394
Heating cables, voltage colors 424-35 p. 393
Heating cables, wallpaper 424-42 p.395
Heating elements, cable separation 424-39 p.393
Heating elements, marking appl. 422-31 p. 387
Heating elements, marking heat 424-29 p.392
Heating elements, resistance 422-28f p. 386
Heating elements, resistance 424-22b p. 391
Heating elements, subdivided 422-28f p.386
Heating equipment branch cts. 424-3a p.388
Heating equipment, metal cover 427-23 page 411
Heating panel set 424-91b page 400
Heating panel set, nailing 424-93b3 page 401
Heating panels, from boxes 8" 424-39 p.393
Heating panels, in concrete 33w 424-98a p. 402
Heating panels, radiant 424-90 page 400
Heating panels, under floor cover 424-99b p.402
Heating tape 427-2 FPN page 409
Heating, disconnect 424-19 p.389
Heating, impedance 426-30 page 407
Heating, induction 427-36 page 412
Heating, induction-dielectric 665 p.739
Heating, pipelines & vessels 427 page 409
Heating, skin effect 426-40 page 407
Heavy-capacity feeders 430-62b p.438
Heavy-duty lampholder 600va 220-3c4 p. 69
Heavy-duty lampholder sockets 210-21a p.60
Heavy-duty track, fixtures 410-103 page 378
Held in free air, opposite polarity T.384-36 p.331
Helically wound flexible conduit 350-2 p. 274
Hermetic comp. disconnect 440-12a1 p.462
Hermetic comp. hp disconnect 440-6a ex.1 p.461
Hermetic refrigerant motor-comp. 440-2 p.459
Hexafluoride gas 325-21 page 233
Hexagonal headed screw, green 410-58b1 p. 372
Hexagonal shape, fuse 240-50c page 116
Hexane 500-3a4 p.496
Hickeys, boxes 370-16b3 page 305
Hickeys, fixtures 410-16d page 363
High bay manufact. bldg. 240-21e pg. 113
High-impedance grd. neutral 250-5b ex.5 p.124
High-impedance shunts 230-82 ex.4 page 101
High-intensity discharge fixtures 410-73f p.374
High-leg, feeder orange color 215-8 p.68
High-pressure spray machines 422-8d3ex. p. 381
High-voltage definition 710-2 p.804
High-voltage fuses 710-21b7 p.810
High-voltage grids spray appl. 516-4a ex. p.562
High-voltage in concrete T.710-4b ex. p. 806
Highest locked-rotor, motors 430-7b 3 p. 418
Highest rated motor, highest FLC 430-17 p.424
Hinged or held captive, lamps 410-82 p.376
Hinged panels or doors 90° 110-16a page 42
Hoists & cranes 610 p.698
Hoists, monorail supports 610-21c p.702
Hoistway door interlock wiring 620-11a p.709
Hoistway riser elevators 620-11a p.709
Hollow spaces, concealed wiring 324-3(1) p.229
Hollow spaces, spread of fire 300-21 p.171
Home economics, demand Note 7 page 76
Hook stick, bus bar disconnect 364-12 p.298
Horizontal runs EMT 348-12 page 272

Horizontal separation 300-21 FPN p.171
Horsepower, controller rating 430-83 p.441
Horsepower, disconnect rated in 430-109 p.447
Horsepower, plug as controller 430-81c p.441
Horsepower, switch rated in 380-15 p.324
Hosedown, motor enclosures T.430-91 p.444
Hospital grade receptacles 517-18b p.572
Hospital lighting demand factor T.220-11 p.72
Hospital, definition 517-3 p.567
Hot tubs & spas 680-40 p.768
Hot tubs, GFCI 680-42 p.770
Hotels & motels receptacles 210-60 p.65
House load Example 4a p.903
House loads 220-32b page 79
Howlers, Class III location 502-14a2 p.524
Human habitation 336-5a page 248
Humidifer branch circuit 422-7 ex. p.380
Hung ceiling 300-22c FPN page 172
Hydrochloric acids, cables 334-4 page 245
Hydromassage bathtub, definition 680-4 p.756
Hydromassage bathtubs GFCI 680-70 p.774
Hydromassage units, cord 422-24 page 384
Hydrotherapeutic tanks 680-62 p.773
Hyperbaric facilities 517-34b4 p.582
Hypobaric facilities 517-34b5 p.582
Hysteresis 300-20b FPN page 171

-I-

Identification of conductors 200 p.49
Identification of disconnect 110-22 p.44
Identification of terminals 200-10b page 51
Ignition systems receptacle 210-52b2 ex.2 p.64
Ignitible concentrations 500-5a p.500
Ignitible flammable gases & vapors 500-5 p.500
Ignitible material, closets 240-24d p.115
Igniting the flammable gas DEF 100 p.31
Ignition of gases, batteries 480-9a p.492
Ignition temp. of gas or vapor 501-8a p.512
Ignition temperature dusts 500-3f p.500
IGS cable, ampacity T.325-14 p.232
IGS cable, gas pressure 20 pounds 325-21 p.233
Illumination for working spaces 110-16e p.43
Illumination, egress lighting 700-16 p.794
Illumination, equipment rooms 110-16d page 43
Image intensifiers, X-ray 517-77 p.596
IMC conduit uses permitted 345-3 page 260
IMC, bends 345-11 page 262
IMC, marking 345-16c page 263
IMC, standard lengths 10' 345-16a page 263
IMC, supports 345-12 page 262
Immersion heaters 422-9 page 381
Immersion heaters 427-2 FPN page 409
Impedance heating 426-30 page 407
Impedance protected, marking 430-7 (14) p.418
Impedance requirements 517-160a FPN1 p.597
Impedance, circuit 110-10 page 39
Impedance, motor windings 430-32c4 p.430
Impedance, transformers T.450-2 FPN p.472
Impeding heat dissipation 310-10 (4) p.179
Impervious metal sheath cable 300-22b p.172
In sight from, A/C disconnect 440-14 p.463
In sight from, controller 430-102a p.446
In sight from, DEF 100 page 30

KEY WORD INDEX

In sight from, duct heater 424-65 p.396
In sight from, motor 430-102b p.446
In sight from, sign 600-6a p.690
In sight from, space heating 424-19a1a p.389
Incandescent lamps, bases 410-53 page 369
Increase heat in transformer 450-3 FPN2 p.472
Increased hazards 230-95 ex.1 p.104
Indicate, its purpose identification 110-22 p.44
Individ. O.C.P. heavy-duty track 410-103 p. 378
Individual b.c. receptacles 210-21b1 p.60
Individual branch circuit 210-3 p.53
Individual holes, boxes 370-17c page 306
Individual open cond., support T.230-51c p.97
Individual transformer definition 450-2 p.471
Individually or in groups, capacitor 460-25c p.488
Individually shielded conductors 300-34 p.173
Indoor antennas box barrier 810-18c p.867
Indoors or outdoors, arresters 280-11 p.157
Induced circulating currents 364-23 p.299
Induction coils, fixed heating 427-36 p.412
Induction generating equip. 705-40 FPN2 p.804
Induction heating coil 665-25 p.741
Induction heating coil over 30v 427-36 p. 412
Induction heating definition 665-2 p.740
Induction, metal raceways 300-20a p.171
Induction-dielectric heating 665 p.739
Inductive current, raceways 300-20a p.171
Inductive effect, minimized 300-20b page 171
Inductive lighting loads 210-22b page 61
Inductive reactance, differences 310-4 FPN p.177
Industrial electric furnaces 250-5b4 ex.1 p.123
Industrial machine 670-2 DEF page 749
Industrial machinery clearance 670-5 p.751
Industrial machinery definition 670-2 p.749
Infrared heating lampholders 422-15b ex. p. 382
Infrared heating lamps 422-15 page 382
Infrared heating, overcurrent 422-28c p. 386
Infrared lamps, an appliance 422-15c page 382
Inherent intermittent duty 675-7 p.752
Inhibitor required, aluminum 110-14 p.40
Inner braids, cords T.400-4 note 3 page 332
Innerduct, plastic 770-5 FPN page 845
Input leads, motor-generator 640-6b p.731
Inserts, floor raceways 358-9 page 290
Inspection, heating cables 424-45 p.395
Instantaneous-trip breaker 430-52c FPN p.435
Instruction manual, untrained person 90-1c p.21
Instructions for installation 110-3b p.38
Instrument circuit protection 384-32 page 330
Instrument transf., grd. cond. 250-125 p.154
Instrument transformers 230-82 ex.4 page 101
Instrument tray cable 310-11b1(h) page 180
Instrument tray cable 310-11b4(f) page 180
Instrumentation tray cable 727-1 DEF page 833
Insulated busbar clearance 384-10 page 327
Insulated fittings 300-4f p.162
Insulated neutral, park trailer 552-55c p. 682
Insulated wires marked or tagged 310-11a p.180
Insulating & structural appl. 110-11 FPN2 p.39
Insulating bushing, threaded 410-30a page 366
Insulating bushings, #4 conductor 300-4f p.162
Insulating joints, bonding 250-81a p.145
Insulating material web, FC cable 363-1 p.294
Insulating mats motor live parts 430-133 p.452

Insulating pipe, bonding 250-81a p.145
Insulation, battery conductors 640-9b p.731
Insulation, lampholder 410-50 page 369
Insulation, voltage stresses 310-6 page 178
Insulators, clean & dry 516-4i page 563
Insulators, strain 225-13 p.86
Integral bonding means 370-3 ex.2 page 302
Integral component of the fixture 210-6c2 p.55
Integral grounding shield 424-99 FPN p. 403
Integral junction box 300-15d p.169
Integral parts of equipment 300-1b page 159
Integrated electrical systems DC 685-12 p.775
Integrated gas spacer cable 325 p.231
Integrity of electrical equipment 110-12c p.39
Interaction nonsinusoidal current 430-2FPN p414
Interactive system, definition 690-2 p.776
Interbuilding cable runs 800-30a FPN4(2) p.855
Intercept lightning 800-30a FPN4(1) p.855
Interchangeable, lighting track 410-105a p. 378
Interconnected power prod. sources 705 p.801
Interconnections, stage 520-53j p.609
Interior metal piping 250-80b page 144
Interior metal water pipe 250-81 ex.2 p.144
Interior metal water piping 250-81a ex. p.144
Interior wiring, SE cable 338-3a p.252
Interlocked metal strip flex. conduit 350-2 p. 274
Intermediate metal conduit 345-3 page 260
Intermediate segments, rheostat 430-82c1 p.441
Intermittent duty, motors T.430-22a ex. p.424
Internal bonding means 370-3 ex.1 page 302
Internal combustion engines 700-12b2 p.793
Internal wiring, factory installed 90-7 p.23
Interpolated, motor hp 430-6a p.417
Interposed in service raceway 250-71a2 p.139
Interrupting rating 110-9 page 39
Interstices between strands 501-5 FPN2 p.506
Interstices, Table C12A Definition Page 1018
Intersystem bonding dwelling 250-71b p.140
Intrinsic safety barrier 504-2 FPN 2 p. 531
Intrinsically safe definition 504-2 p.532
Intrinsically safe circuit DEF 504-2 p.531
Intrinsically safe connections 504-2 FPN 1 p.531
Intrinsically safe systems 504 p.531
Introduce noise of data errors 250-21d p.126
Inverter 690-2 DEF page 776
Iron, cords 422-8a p.380
Iron, flat iron 422-13 p.382
Irradiance enhancement 690-52 FPN p.782
Irradiated, solar 690-18 FPN p.780
Irradiation, X-ray 660-23b p.738
Irreversible compression con. 250-81 ex.1 p.144
Irreversible compression con. 250-91 ex.3 p.147
Irrigation cable 675-4 p.751
Irrigation machines, ground. 675-13 p.754
Irrigation, lightning protection 675-15 p.754
Island counter top receptacles 210-52c2 p.64
Isolated by elevation 8' 430-132c p.452
Isolated conductors, color 517-160a5 p.597
Isolated ground receptacles 410-56c page 370
Isolated grounding 250-74 ex.4 p.141
Isolated phase installations 300-5i ex.2 p.165
Isolated power system, definition 517-3 p.567
Isolating switches over 600v 230-204d p.105
Isolating switches, knife 380-13a page 323

KEY WORD INDEX pg - 21

Isolation transformer 517-160 FPN1 p.597
Isolation transformer, definition 517-3 p.567
Isolation transformer, heating 427-27 p.412

-J-

Jacks, communication circuits 800-4 ex.1 p.852
Jars, battery cells 480-5b p.491
Joined, mechanically/electrically 110-14b p. 40
Joints, cord tension 400-10 p.347
Joints, expansion 300-7b p.166
Joints, festoon wiring staggered 520-65 p.611
Joints, ground without splice 250-91a p.146
Joints, PVC conduit 347-6 page 269
Joists, bored holes for cable 300-4a1 p.160
Jumpers, bonding 250-79 p.142
Junction box from motor 430-145b p.453
Junction boxes 370-28 p.309
Junction boxes, accessible 370-29 p.310
Junior hard service cords T.400-4 page 337

-K-

Keying, electrolytic cells 665-67 p.743
Kitchen equip., feeder demand T.220-20 p.76
Kitchen hood supply exhaust 517-34a5 page 581
Kitchen hood supply exhaust 517-43a4 page 587
Kitchen sink, GFCI 210-8a7 page 57
Kitchen waste disposers, cord 422-8d1 p.381
Kitchen, small appliances b.c. 210-52b1 p.63
Kitchen, small appliances feeder 220-16a p.73
Kitchen, two 20 amp circuits 220-4b p.71
Knife switches 380-6 page 321
Knife switches 600 volt 380-16 p.324
Knife switches double throw 380-6b p.321
Knife switches exposed blades 384-34 p.330
Knife switches height 6' 7" 380-8a p.322
Knife switches over 1200 amp 380-13a p.323
Knife switches renewable contacts 380-16 p.324
Knob & tube wiring, extensions 324-3 p.229
Knob and tube wiring 324-3 page 229
Knob and tube wiring, box 300-15b page 168
Knob and tube wiring, splices 324-12 p. 231
Knobs, glass or porcelain 225-12 p.86
Knobs, open wiring support 320-7 p.226
Knockouts, close open. cabinets 373-4 p.313
Knockouts, closed openings boxes 370-18 p.306
Knotting the cord 400-10 FPN page 347
KVA, capacitor marking 460-12 p.488
KVA, vault 112 1/2 450-26 ex.1 p.481

-L-

Labeled equipment 110-3b page 38
Ladders, clearance 230-9 p.91
Ladders, facilitate the raising 225-19e p.88
Laid on the floor, conductors 305-4c p.174
Lamp bases, mogul 410-53 page 369
Lamp cord T.400-4 page 333
Lamp replacement fixtures 410-82 page 376
Lamp tie wires, fixtures 410-19b ex. page 364
Lamp wattage marking 410-70 page 374
Lamp, electric-discharge 410-54a page 370
Lampholder, insulation 410-50 page 369

Lampholder, lead wire 410-51 page 369
Lampholder, screw shell ground 410-23 p. 364
Lampholder, switching device 410-48 page 369
Lampholders 410 page 358
Lampholders, admedium type 210-21a p.60
Lampholders, candelabra-base 410-27b p.365
Lampholders, cleat-type 410-3 ex. page 358
Lampholders, double-pole switched 410-48 p.369
Lampholders, dressing rooms 520-71 p.613
Lampholders, heavy-duty 210-21a page 60
Lampholders, infrared 422-15b ex. page 382
Lampholders, lead wires #14 410-51 p.369
Lampholders, less than 50v 720-5 p.819
Lampholders, metal to cord/bush. 410-30a p.366
Lampholders, outdoor 225-24 page 88
Lampholders, over combustible area 410-6 p.359
Lampholders, pendant 410-27a page 365
Lampholders, porcelain const. 410-72 p. 374
Lampholders, screw-shell type 410-47 p.369
Lampholders, switch center contact 410-52 p.369
Lampholders, unswitched type 410-6 page 359
Lampholders, viewing tables 530-41 p.621
Lamps, back stage 520-47 page 606
Lamps, bare bulbs 520-47 page 606
Lamps, film storage vaults 530-51 p.621
Lamps, outdoor lighting 225-25 p.88
Lanterns, stage 520-65 p.611
Lateral displacement, fixtures 501-9a3 p.513
Lateral, service definition 100 p.34
Laundries, wet location 300-6c page 166
Laundry branch circuit 20 amp 220-4c p. 71
Laundry facilities Example 4a p.903
Laundry on premises Example 4a p.903
Laundry outlet 6' from appliance 210-50c p.63
Laundry, feeder load 1500va 220-16b p. 74
Lead covered conductors 640-3 ex.1&2 p.730
Lead covered insulated conductors 310-8a p.179
Lead sheath cable 710-4b ex.2 p.805
Lead sheath, field bends T.346-10 ex. page 266
Lead wire, heating cable 424-35 p.393
Lead wire, lampholder 410-51 page 369
Lead-ins, antennas 810-18c p.867
Leads, of motors, enclosing 350-10a1 page 275
Leads to motor-generator 640-6b p.731
Leads to rotary converter 640-6b p.731
Leads, storage batteries 640-9b p.731
Leakage from piping systems 384-4 ex. p.326
Legally req. standby systems, power 701-11 p.798
Legally required standby systems 701 p.796
Length of cord, compactors 422-8d2a page 381
Length of cord, dishwasher 422-8d2a page 381
Length of cord, disposers 422-8d1a page 381
Length of free conductor 6" 300-14 p.168
Less than 1000v grounded 250-23b page 127
Letter suffixes conductors 310-11c p.181
Lieu of box 300-15d p.169
Life safety branch, definition 517-3 p.568
Lifting handles, over 600 volts 710-24i p.813
Light fixture lens, pool 680-20a3 p.761
Light fixtures, over pool 680-6b p.758
Lighting fixtures 410 page 358
Lighting fixtures, over tub 410-4d p.359
Lighting fixtures, stage 520-42 p.604
Lighting outlet, attic 210-70a page 66

KEY WORD INDEX

Lighting outlet, auto control 210-70a ex.2 p.66
Lighting outlet, basement 210-70a page 66
Lighting outlet, guest rooms 210-70b page 66
Lighting outlet, heat-AC 210-70c page 66
Lighting outlet, stairway 210-70a page 66
Lighting outlet, underfloor space 210-70a p.66
Lighting outlet, utility room 210-70a page 66
Lighting outlet, vehicle door 210-70a p.66
Lighting outlets 210-70 page 66
Lighting systems 30 volts or less 411-1 page 378
Lighting track 410-100 page 377
Lighting track b.c. calculation 410-102 p.378
Lighting track conductors 410-105a page 378
Lighting track heavy-duty 410-103 p.378
Lighting track interchangeable 410-105a p. 378
Lighting track RMS voltage 410-101c p.377
Lighting, border lights 520-44 p.605
Lighting, Christmas tree 410-27b ex. page 365
Lighting, clothes closets 410-8a p.359
Lighting, cove 410-9 page 361
Lighting, egress 700-16 p.794
Lighting, emergency 700-12a p.793
Lighting, exit 700-16 p.794
Lighting, festoon 225-6b p.84
Lighting, means of egress 517-32a p.579
Lighting, proscenium 520-44a p.605
Lighting, railway conductors 110-19 p.44
Lighting, show window 200va 220-12 p.73
Lighting, simulating 520-66 p.612
Lighting, track construction 410-105 p.378
Lighting, trees 410-16h p.363
Lighting, vehicle lanes 12' 511-7b p.541
Lightning arrester conductor #6 280-23 p.158
Lightning conductors, CATV 820-10f3 p.872
Lightning prot., irrigation mach. 675-15 p.754
Lightning rods, spacing 6' 250-46 page 135
Lightning rods, use 250-86 page 146
Light-emitting diodes 504-2 FPN page 532
Lights 30 volts or less 411-1 page 378
Lights, dressing rooms 520-73 p.613
Limit controls, duct heater 424-64 p. 396
Limit switch for hoist 610-55 p.706
Limitation, lighting voltage 210-6a1 p.54
Limitation, voltage 300-2a p.159
Limited care facility 517-3 p.568
Limited flexibilty motor connect. 501-4b p.505
Limited smoke T.400-4 note 9 page 344
Limited-smoke characteristics 333-22 p.245
Limited-smoke LS marking 310-11d FPN p.181
Limited-smoke marking 400-6b FPN p.347
Limited-smoke marking 402-9c FPN p.357
Limited-smoke markings 347-17 FPN p.270
Limiter, fusible connector 450-6a3 p.477
Limiting the number of circuits 90-8b p.23
Line frequency in converter 665-66 p.743
Line isolation monitor 517-160b1 p.597
Line isolation monitor 517-19e ex. p. 573
Line isolation monitor, definition 517-3 p.568
Line monitor alarm 517-160b p.597
Line to neutral voltage 240-60a ex. p.117
Linear feet, show window recpt. 210-62 p.65
Linear foot, heating cable 16 1/2w 424-44a p.395
Linoleum, underfloor raceway 354-2 p.286
Liquid spillage FCC cable 328-11 page 235

Liquid that will burn, Class I 501-2a1 p.503
Liquids, exposed to 110-11 page 39
Liquidtight conduit, bends 351-10 page 278
Liquidtight conduit, uses 351- page 277
Liquidtight flexible conduit 351 page 277
Liquidtight flexible conduit 553-7b p.685
Liquidtight, nonmetallic flex 351-22 page 278
Listed baseboard heaters 210-52a FPN p.63
Listed Christmas tree lighting 410-27b ex. p.365
Listed devices closed-loop systems 240-20c p.111
Listed equip. double insulated 250-42 ex.4 p.132
Listed equip. double insulated 250-45 ex.3 p.134
Listed equipment 110-3b page 38
Listed incandescent fixtures 210-6c2 p.55
Listed potting compound 680-20b1 p.761
Listed, signs 600-3 p.689
Live parts, air space cabinet 373-11a3 p.317
Live parts, appliances 422-2 ex. page 379
Live parts, exposed Table 110-16a page 42
Live parts, fixtures 410-3 p.358
Live parts, guarding 110-17a page 43
Live parts, motors 430-132 p.451
Live parts, porcelain fixture 410-46 page 369
Live parts, switchboards 110-16a1 p.42
Live parts, unguarded over 1000v T.110-34e p.48
Live vegetation 225-26 page 89
Livestock facility, potential 547-8c FPN p.630
Livestock is housed 250-24b ex.2d page 128
Load additions existing dwellings 220-3d1 p.71
Load balancing of circuits 220-4d page 71
Load diversity Note 8 FPN p.196
Load diversity Table B-310-11 page 935
Load end of service drop 250-23a page 126
Load interrupters over 600v 710-21e p.811
Load management, services 230-94 ex.3 p.103
Load pickup, emergency systems 700-5b p.790
Load shedding, emergency systems 700-5b p.790
Loads on emergency branch cir. 700-15 p.794
Loads, dissimilar 220-21 p.76
Loads, existing dwelling 220-3d1 page 71
Localizing a fault condition 240-12 FPN p. 111
Location boards, motion pictures 530-18c p.619
Location of lamps, outdoor lts. 225-25 p.88
Location overload units T.430-37 p.432
Locked in open position 430-102 ex.1 p.446
Locked-rotor curr. small motor 430-110c3 p.449
Locked-rotor current rating 440-41b p.466
Locknuts, double bonding 501-16a p.516
Longer motor acceleration 430-34 FPN p.431
Loop wiring, cellular raceways 356-6 page 288
Loop wiring, underfloor raceways 354-6 p.287
Loop, switch white conductor 200-7 ex.2 p.50
Loose, rolled, or foamed insulation 324-4 p.230
Loudspeakers, protection 640-13 p.732
Louvers, vault door transformer 450-45c p.483
Low ambient condition 500-3 FPN 2 p.496
Low impedance grounding 250-51 p.135
Low impedance path 250-1 FPN1 p.120
Low leakage insulation 517-160 FPN 1 p.597
Low-density insulation fiberboard 410-76b p.375
Low-frequency AC 665-28 p.741
Low-smoke prod. characteristics 300-22c p.172
Low-smoke producing cable 725-71a FPN p.831
Low-voltage equipment 10 volts 517-64a1 p.593

KEY WORD INDEX pg - 23

Low-voltage OCP T. 552-10e1 page 670
Low-voltage systems 720 p.819
Lower threshold value 517-160b1 ex. p.598
Lowest temperature rating 110-14c p.41
Lubrication and service rooms 511-3b ex. p.539
Lugs, electrode connections 250-115 p.152
Lugs, grounding 250-113 p.152
Lugs, service conductors 230-81 p.101
Lugs, terminal connections 110-14a p.40
Luminaire, definition 410-1 FPN page 358

-M-

Machine tool nameplate 670-3a p.750
Machine tool wire T.310-13 p.183
Machine tools 670 p.749
Machines, curtain 520-48 p. 606
Main bonding jumper 250-79a p.142
Main, service taps 230-46 ex #2 p.96
Maintain concentricity note 5 T.400-4 p.332
Maintenance, proper 90-1b p.21
Major diam.of the ellipse Table 1 note 9 p.879
Make-or-break contacts, windings 501-3b3 p.504
Malfunctioning receptacle 555-4 FPN p.687
Malleable iron box 370-40b page 311
Malleable iron, fixture studs 410-16d p.363
Mandatory rules shall 90-5 page 23
Manipulating, gaging, measuring 670-2 p.749
Manual override 210-70a ex.3 p.66
Manufactured build., service 545-6 ex. p.627
Manufactured buildings 545 p.626
Manufactured home 550-2 DEF page 631
Manufactured phase converter 455-2 page 484
Manufactured wiring systems 604 p.695
Manufacturer's name on product 110-21 p. 44
Manufacturer's name on product 310-11a3 p.180
Marinas & boatyards 555 p.686
Marking appliances 422-30a page 387
Marking of heating cables 424-35 p.393
Marking of heating elements 422-31 p. 387
Marking tubing over 1000v 410-91 page 377
Marking, motor control centers 430-98 p.446
Marking, phase converter 455-4 p.484
Marking, service equipment 230-66 p.99
Marquees, fixture location 410-4a p.358
Marquees, receptacles 410-57a page 371
Mast weatherhead, mobile homes 550-5i p.634
Mast, service 230-28 p.93
Master handle, breaker 230-71b page 99
Material handling magnet cir. 240-3a p. 108
Material that envelops the cond. 324-4 p.230
Mats or platforms for motors 430-133 p.452
Mats, rubber for switchboards 250-123c p.154
MC cable definition 334-1 page 245
MC cable, as aerial cable 334-10e page 246
MC cable, bending radius 334-11 page 246
Means of egress illumination 517-42a p.586
Means of egress, lighting 517-32a p.579
Mechanical continuity raceways 300-12 p.167
Mechanical execution of work 110-12 p. 39
Mechanical execution of work 800-6 p.853
Mechanical protect., buried cables 300-5d p.162
Mechanical protection, conductors 300-4 p.160
Mechanical protection, romex 336-6b p. 249

Mechanical protection, service 230-50 p. 96
Mechanical strength for parts 110-3a2 p. 38
Mechanical ventilation 511-3 ex. p. 539
Medium base incandescent lamps 410-53 p.369
Medium density polyethylene 325-22 page 233
Medium voltage cable definition 326-1 p.233
Medium-drawn copper span 35' 810-11 ex. p.866
Melting point or trip setting 230-208 p.106
Melting point or trip setting 240-100 p.119
Melting point or trip setting 240-101 p.119
Mercury vapor, emergency lights 700-16 p.794
Messenger cable, physical damage 321-4 p.229
Messenger cable, spans over 40' 225-13 p.86
Messenger wire, festoon supports 225-13 p.86
Metal air ducts, bonding 250-80 FPN p.144
Metal canopies, fixtures 410-38b page 368
Metal car frames, grounded 250-58b p.137
Metal clad cable definition 334-1 p.245
Metal cover, electric heat 427-23 page 411
Metal dusts, transformers 502-2a3 p.513
Metal elbow 250-32 ex. p.131
Metal elbow 250-33 ex.4 p.132
Metal embedded in floor 424-44d page 395
Metal enclosed busways temp. rise 364-23 p.299
Metal faceplates, thickness 380-9 p.322
Metal faceplates, thickness 410-56d p.370
Metal frame bldg. ground 250-58a p.137
Metal halide, emergency lights 700-16 p.794
Metal hood, switchboard 520-24 p.602
Metal lampholder, insulating bush.410-30a p.366
Metal oxide arrester 280-4 FPN2 page 157
Metal plugs or plates, boxes 370-18 page 306
Metal poles handhole 410-15b1 p.362
Metal poles support fixtures 410-15b p.362
Metal sheaves, grounding 250-58b page 137
Metal shield connections definition 328-2 p. 235
Metal studs, bushing for cable 300-4b1 p.161
Metal surface raceways 352 p.280
Metal underground water pipe 250-81a p.144
Metal water piping bonding 250-80a p. 143
Metal well casing, bonding 250-81a page 144
Metal well casings 250-431 p.133
Metal, molten arcing parts 110-18 p. 44
Metal-to-metal bearing surfaces 610-61 p.706
Metallic interconnections 280-24a p.158
Metallic power service raceway 800-40b1 p.858
Metallically isolated water pipes 250-80 ex p.143
Metallized foil shield, signal 725-71e p.831
Meter loops, cabinets 373-11c p. 317
Meter socket enclosures 230-66 p. 99
Metering equip.connections 230-46 ex.1 p. 95
Meters, ahead of main 230-82 ex.3 p.101
Methane 500-3a4 p.496
Methanol 500-3a4 p.496
Methods of construction 310-4 FPN p.177
Metric trade numerical dimen. 345-6 FPN p.261
Metric trade numerical dimen. 346-5 FPN p.264
Metric trade numerical dimen. 348-5 FPN p.271
Metric units of measurement 90-9 page 23
Mezzanine, access 110-33b page 46
MG set Example 9 p. 910
MI cable nonmagnetic sheath 300-3b ex.1 p.160
MI cable, bends 330-13 page 239
MI cable, definition 330-1 p.238

KEY WORD INDEX

MI cable, end seal Note 6 page 195
MI cable, fittings 330-14 page 239
MI cable, ground marking 210-5 ex.1 p.54
MI cable, outer sheath 330-22 page 239
MI cable, sealing 501-5 p.506
MI cable, solid copper 330-20 page 239
MI cable, temperature limitations Note 6 p.195
MI cable, terminal seals 330-15 page 239
MI cable, uses permitted 330-3 p.238
Mineral insulated cable definition 330-1 p.238
Mineral insulated cable, ground 210-5 ex.1 p.54
Mines, underground 90-2b2 page 21
Minimize the effects from a short 90-8b p.23
Minimum cover T.300-5 page 163
Minimum number branch circuits p.897
Minimum radius conduit bend T.346-10 p.265
Minor relative motions 545-13 p.628
Minus 10° C (plus 14° F) 310-13 FPN p.182
Minus 10° C (plus 14° F) 402-3 FPN page 350
Mirror frames 680-41d4 ex. p.769
Mobile homes & parks 550 p.631
Mobile homes, appliances 550-7c p.636
Mobile homes, bonding 550-11c p.640
Mobile homes, branch circuit 550-7 p.635
Mobile homes, calculations 550-13 p.641
Mobile homes, demand factors 550-22 p.643
Mobile homes, disconnect 550-6a p.634
Mobile homes, disconnect height 550-23e p.645
Mobile homes, distribution system 550-21 p.643
Mobile homes, feeder 550-24 p.645
Mobile homes, GFCI 550-8b p.636
Mobile homes, heat tape outlet 550-8g p.637
Mobile homes, insul. neutral 550-11a1 p.640
Mobile homes, lighting circuits 550-7a p.635
Mobile homes, portable appl. 550-2 FPN p.631
Mobile homes, power supply 550-5a p.632
Mobile homes, receptacles 550-8a p.636
Mobile homes, service 550-21 p.643
Mobile homes, service rating 550-23b p.644
Mobile homes, supply cord length 550-5d p.633
Mobile homes, testing insulation 550-12 p.641
Mobile homes, weatherhead 550-5i p.634
Mobile homes, wiring methods 550-10 p.638
Mobile shovels over 600v 710-41a p.815
Mobile x-ray definition 517-3 p.569
Module, solar definition 690-2 p.776
Modules, buildings 545-13 p.628
Mogul base, incandescent 410-53 p.369
Mogul-base lampholders 410-25a page 364
Moisture-resistant 339-1a page 253
Moisture-resistant NMC cable 336-30b2 p. 250
Monitor hazard current 517-160b2 FPN p.598
Monitor hazard current, definition 517-3 p.567
Monitoring 240-12 FPN page 111
Monorail hoists 610-21c p.702
Monorail track 610-21f p.703
Motel conference room 100 persons 518-2 p.598
Motion pict. proj., work space 540-12 p.625
Motion picture projectors 540 p.624
Motion picture studios feeders 530-18b p.619
Motor control center grounding 430-96 p.445
Motor control centers 430-92 p.444
Motor control units 430-98b p.446
Motor-comp. controller 440-41b p.466

Motor-comp. plug rating 440-55b p.468
Motor-compressor time delay 440-54b p.468
Motor, disconnect Design E 430-109 ex. 1 p.447
Motor-generator equipment 665-40 p.741
Motor-generator leads 640-6b p.731
Motor leads, enclosing 350-10a1 page 275
Motor terminal,Flex metal conduit 430-145b p.453
Motors 430 page 414
Motors, accidental ground 430-73 p.440
Motors, accidental starting 430-73 p.440
Motors, adjustable speed drive 430-2 p.414
Motors, armature shunt resistors 430-29 p.427
Motors, attachment plug rating 430-42c p.433
Motors, automatic restarting 430-43 p.433
Motors, automatic starting 430-35b p.431
Motors, autotransformer 430-82b p.441
Motors, auxiliary leads T.430-12b p.422
Motors, branch circuit 430-22a p.424
Motors, branch circuit O.C.P. T.430-152 p.458
Motors, brush rigging 430-132c ex. p.452
Motors, bushing 430-13 p.423
Motors, can be started kva 430-7b (1) p. 418
Motors, capacitors 430-27 p.427
Motors, clock 430-32c(4) FPN p.430
Motors, collector rings 430-14b p.423
Motors, combination load 430-25 p.426
Motors, commutators 430-14b p.423
Motors, conductor fuses 430-36 p.431
Motors, contact arm 430-82c1 p.441
Motors, continuous duty T.430-22a ex. p.425
Motors, control circuit torque 430-9c p.420
Motors, control circuits 430-71 p.438
Motors, control transformer 430-74b p.441
Motors, controllers 430-81 p.441
Motors, controllers rating T. 430-91 page 444
Motors, cord & plug connected 430-42c p.433
Motors, DC contant voltage 430-29 p.427
Motors, DC fractional 430-7a(12) p.418
Motors, Design E 430-52c3 ex. 1 p. 435
Motors, Design E controller 430-83a ex. 1 p. 442
Motors, disconnect 430-101 p.446
Motors, disconnect 100 hp 430-109 ex.4 p.447
Motors, disconnect amp rating 430-110a p.448
Motors, disconnect rated in hp 430-109 p.447
Motors, dust accumulation 430-16 p.423
Motors, duty-cycle T.430-22a ex. p.425
Motors, dynamic braking 430-29 p.427
Motors, electrical resonance 430-2 FPN p. 414
Motors, enclosed pos. pressure vent. 501-8a p.512
Motors, escalator 620-61b2 p.719
Motors, exciting fields T.430-12b p.422
Motors, exposure to dust 430-16 p.423
Motors, factory connections 430-12d p.421
Motors, fan cooled 503-6 p.527
Motors, fault-current prot. 430-125c p.451
Motors, feeder conductors sizing 430-24 p.426
Motors, feeder demand 430-26 p.427
Motors, feeder protection sizing 430-62 p.437
Motors, feeder taps 430-28 p.427
Motors, future additions 430-62b p.438
Motors, grounding 430-141 p.452
Motors, guards for attendants 430-133 p.452
Motors, hazard to persons 430-125a ex. p.450
Motors, heater sizing maximum 430-34 p.430

KEY WORD INDEX pg - 25

Motors, heater sizing minimum 430-32 p.428
Motors, heavy-capacity feeders 430-62b p.438
Motors, hi-voltage O.C.P. 430-125c1a p.451
Motors, highest locked-rotor 430-7b (3) p. 418
Motors, highest rated 430-17 p.424
Motors, large-capacity inst. 430-62b p.438
Motors, letter code table T.430-7b p.419
Motors, limited flexibility 501-4b p.505
Motors, live parts 430-132 p.451
Motors, locked in open 430-102b ex.1 p.446
Motors, locked-rotor 430-7b page 418
Motors, locked-rotor current 430-110c3 p.449
Motors, marked diameter 430-7a(12) p.418
Motors, mats or platforms 430-133 p.452
Motors, molded case switch 430-109 p.447
Motors, multispeed 430-7b (1) page 418
Motors, nameplate 430-7a p.417
Motors, next higher size 430-52a ex.1 p.434
Motors, nonautomatically 430-35a p.431
Motors, nonburning housings 430-12aex. p.421
Motors, nonsinusoidal currents 430-2 FPN p.414
Motors, nonventilated enclosed 503-6 p.527
Motors, occasional prolonged sub. T.430-91 p. 444
Motors, oil switch 430-111c p.449
Motors, open 430-14b p.423
Motors, orderly shutdown 430-44 p.433
Motors, over 600 volts 430-121 p.450
Motors, overload protection 430-55 p.437
Motors, overload sizing 430-32 p.428
Motors, overload units T.430-37 p.432
Motors, PF correction capacitors 430-2FPN p.414
Motors, pipe-ventilated 430-16 FPN p.424
Motors, pipe-ventilated 503-6 p.527
Motors, portable 1/3 hp 430-81c p.441
Motors, power & light loads 430-63 p.438
Motors, prot. live parts 430-131 p.451
Motors, rating of plug 430-42c p.433
Motors, rectifier voltage 430-18 p.424
Motors, resistor duty T.430-23c p.426
Motors, rheostats 430-82c p.441
Motors, second. cir. wound-rotor 430-32d p.430
Motors, separation of junct. box 430-145b p.453
Motors, service factor 430-32a1 p.429
Motors, shaded-pole 440-6b p.461
Motors, shunting 430-35 p.431
Motors, simultaneously disc. 430-125c1a p.451
Motors, single-phase F.L.C. T.430-148 p.454
Motors, sparks 430-14b p.423
Motors, split-capacitor 440-6b p.461
Motors, split-phase 430-32 FPN p.430
Motors, stage curtains 520-48 p.606
Motors, stationary 2 hp 430-109 ex.2 p.447
Motors, stationary frame grd. 430-142 p.452
Motors, temperature rise 430-32a1 p.428
Motors, terminal housings 430-12a p.421
Motors, terminal spacing T.430-12c1 p.423
Motors, three-phase F.L.C. T.430-150 p.456
Motors, torque 430-7c page 419
Motors, torque requirements 430-9c p. 420
Motors, totally enclosed 501-8a p.512
Motors, usable volumes T.430-12c2 p.423
Motors, ultimate trip-current 430-126 p. 451
Motors, ventilation 430-14a p.423
Motors, voltage rating 430-83b p.442

Motors, winding impedance 430-32c4 p.430
Motors, wire bending space T.430-10b p.421
Motors, wire-to-wire connect. T.430-12b p.422
Motors, wooden floors 430-14b p.423
Motors, wound-rotor 430-23a p.425
Motors, wound-rotor secondaries 430-32d p.430
Motors, wye start-delta run 430-22a page 425
Motors. D.C. F.L.C. T.430-147 p.453
Mounting of boxes 370-23a p.307
Mounting of equipment 110-13a p.40
Mounting of fixtures 410-15 p.362
Mounting of switches 380-10 p.323
Mounting plane of the fixture 410-46 page 369
Mounting screws, glass lamps 410-19 ex. p.364
Mounting yoke 380-10b page 323
Moving walks 620 p.706
Multi-car installations, elevators 620-52a p.718
Multicircuit track fixtures 410-100 page 377
Multiconductor portable cables 400-30 p.349
Multifamily dwelling, optional T.220-32 p.79
Multioutlet assemblies 220-3c ex.1 page 70
Multioutlet assembly dry location 353-2 p.285
Multiple DC voltages 690-8c p.778
Multiple electrodes spacing 250-83-4 p.145-46
Multiple raceways grounding 250-95 p.149
Multiplexed single conductor cables 300-34 p.173
Multipurpose cables (MP) 800-51g p.861
Multisection enclosures 501-8b FPN2 p.513
Multiwire branch circuits 210-4 p.53
Multiwire branch circuits 501-18 p.517
MV cable 35kv, uses permitted 326-3 p.233
MV cable definition 326-1 page 233

-N-

Nails to mount knobs 10 penny 320-7 p.226
Naphtha 500-3a4 p.496
Natural gas 500-3a4 p.496
Natural gas rated pipe 325-22 page 233
Navigable water, wiring 555-8 p.688
Neat and workmanlike manner 110-12 p. 39
Neon Tubing 600-2 DEF page 689
Neutral can be reduced 551-72 p.665
Neutral conductor Note 10a,b,c page 197
Neutral conductor DC generators 445-5 p. 470
Neutral conductor routing 250-27d page 131
Neutral conductor, cords 400-5 page 344
Neutral fully insulated 250-27b page 131
Neutral grounding impedance 250-27c p. 131
Neutral lead, overcurrent coil 530-63 p.622
Neutral, ampacity solar 690-62 p.782
Neutral, autotrn. phase current 450-5 FPN p.475
Neutral, bare service 230-22 ex. p.91
Neutral, boiler over 600v 710-72e p.818
Neutral, bonding to service 250-72a p.140
Neutral, busway 364-24 page 299
Neutral, carry the unbalance Note 10a p.197
Neutral, clothes dryer feeder 70% 220-22 p.76
Neutral, cook. equip. B.C. 210-19b ex 2 p.59
Neutral, cook. equip. feeder 70% 220-22 p. 76
Neutral, feeder load 220-22 page 76
Neutral, feeders 215-4 p.67
Neutral, insul., park trailers 552-55c p. 682
Neutral, insul., mobile homes 550-11a1 p.640

KEY WORD INDEX

Neutral, min. insulation level 250-152a p.155
Neutral, minimum service size 250-23b p.127
Neutral, outside wiring 225-7b p.84
Neutral, portable switchboards 520-53o p.610
Neutral, reduction feeder 220-22 p.76
Neutral, solid-state dimmers 520-27a1 p.603
Neutral, solidly grounded 250-152 p.155
Next higher O.C.P. device 240-3 b page 108
NFPA Regulations Gov. Committe 90-6 FPN p.23
Night club lighting dimmer 520-25a p.602
Nipple, 24" & 60% fill Chapter 9 note 4 p.879
Nipple, derating factor Note 8 ex.3 page 196
NM cable, protection 336-6c page 249
NM cable, supports 336-18 page 250
NM cable, uses 336-4 page 248
NMC cable, construction 336-30a page 250
NMC cable, definition 336-2 p.248
No appliances and no lamps 700-15 p.794
No equipment ground 210-7d3b p.56
No heating due to hysteresis 300-20b FPN p.171
No point along the floor line 6' 210-52a p.63
No-niche fixture 680-20d p.762
Nominal battery voltage definition 480-2 p.490
Nonautomatically started motor 430-35a p.431
Nonbridge structures, clearance 225-19b p.88
Nonbuilding 225-19b p.88
Noncapacitive load, signal 725-21a2 p.822
Noncarbon arc discharge lamp 530-17b p.619
Noncinder concrete 345-3c page 261
Noncohesive granulated soil 370-29 ex. p. 310
Noncoincident loads, omit smaller 220-21 p. 76
Noncombustible cases, lamps 410-54a p.370
Nonconductive coatings, removed 250-118 p.153
Nonconductive paint, bonding 250-75 p.141
Nonconductive rope operator 668-32b2 p.748
Nonconductive, optical fiber 770-4a p.845
Nonferrous metal conduit 346-1c p. 263
Nonferrous metal, electrodes 250-83d p. 146
Nonfire-rated floor/ceiling 300-11
Nongrounding type receptacles 210-7d(3) p.56
Nonhazardous location 516-2d p.561
Nonheating leads, cables length 424-34 p.392
Nonhygroscopic, irrigation cable 675-4a p.751
Nonincendive circuit DEF 100 p.31 .
Nonincendive circuits 501-14b1 ex.c p.516
Nonincendive circuits 501-3b1 ex.c p.504
Nonincendive circuits 502-4b ex. p.519
Nonincendive circuits, Class I 501-3b1c p.504
Noninsulated busbar clearance 384-10 p. 327
Noninterchangeable cartridge fuse 240-60b p.118
Noninterchangeable S fuse 240-53b p.117
Nonlinear currents 450-3 FPN2 p.472
Nonlinear loads 400-5 p.345
Nonlinear loads 210-4a FPN p.53
Nonlinear loads 220-22 p.77
Nonlinear loads DEF 100 p.31
Nonlinear loads Note 10c p.197
Nonmagnetic sheath MI cable 300-3b ex.1 p.160
Nonmedical or nondental, X-ray 660-1 p.736
Nonmetallic auxiliary gutters 374-9e2 p. 320
Nonmetallic boxes, supports 370-23 p.307
Nonmetallic boxes, uses 370-3 page 302
Nonmetallic extensions definition 342-1 p.255
Nonmetallic fillers, cables T.400-4 note 5 p.332

Nonmetallic frames, space heat 424-44c p. 395
Nonmetallic surface extensions 342-7 p. 257
Nonmetallic wireways 362-14 p.292
Nonmetallic wiring, bulk plant 515-5c p.551
Nonmetallic-sheathed cable 336 page 248
Nonmotor appliance overcurrent 422-28e p. 386
Nonorderly shutdown 230-95 ex.1 p.104
Nonplasticized PVC 331-1 page 240
Nonpolarized attachment plug 422-23 ex. p. 384
Nonpower-limited fire circuit 340-4 p.255
Nonpower-limited fire signaling 760-21 p.835
Nonremovable S fuse adapters 240-54c p.117
Nonshielded conductors 300-3c1 p. 160
Nonshielded conductors 310-6 ex. page 178
Nonshielded high-voltage cables 710-4b p.805
Nonsinusoidal currents, motors 430-2 FPN p. 414
Nontamperable circuit breaker 240-82 p.118
Nontamperable S fuse 240-54d p.117
Nonwicking filter, machines 675-4 p.751
Notches in wood 300-4a2 page 161
NPLFP cable 760-30b1 p.837
NPLFP cable, fire signal 760-31d p.838
NPT conduit cutting die 500-2 p.493
Number of bends in one run 346-11 page 265
Number of circuits in enclosures 90-8b page 23
Number of cond. in conduit 640-3 ex.1 &2 p.730
Number of conductors in conduit T. 3c p.950
Number of wires in conduit T. C3-12a p.941-1018
Nurses' stations, definition 517-3 p.568
Nursing home, definition 517-3 p.568

-O-

Obelisk † T.310-16 page 191
Objectionable current, grounding 250-21a p.125
Oblique angle, ground rod 250-83c3 p. 146
Occasional prolonged submersion T. 430-91 p.444
Occasional temp. submersion T.430-91.p.444
Occupy same raceway, diff. volts 725-26b p.823
Occupancy sensors 210-70a ex.3 p.61
OCP industrial machinery 670-1 page 749
Office furnishings, recpt. outlet 605-5c p.697
Office, unknown receptacles T.220-3b p.70
Offset, enclosures 373-6b FPN p. 315
OFN & OFC cables 770-51d p.847
OFNP & OFCP cables 770-51a p.846
OFNR & OFCR cables 770-51b p.847
Oil circuit breakers 710-21a3 p.808
Oil switch service disconnect 230-204a p.105
Oil switch, motor 430-111c p.449
Oil-filled cutouts 710-21d p.810
Oil-fired central heating 550-5a ex.1 p.633
Oily area, conductors T. 310-13 page 183
One conductor diameter, spacing 365-3d p. 301
One conductor per ground clamp 250-115 p.152
One neutral several circuits 215-4a p.67
One outlet for sign-outline light.600-5a p.690
One shot bender 346-10 ex. page 265
One wire under screw terminal 110-14a p.40
Open circuit voltage dwelling 410-80a p.376
Open motors with commutators 430-14b p.423
Open or partially enclosed lamps 410-8c p.361
Open porches, canopies, marquees 410-57a p.371
Open service conductors supports T.230-51c p.97

KEY WORD INDEX pg - 27

Open spraying classification 516-2b1 p.556
Open wire systems on insulators 320-16 p.228
Open wire systems on insulators 320-6 p.226
Open wire systems on insulators 380-10a p.323
Open wiring, accessible attics 324-11a p.230
Open wiring, conductor separation 225-14c p.86
Open wiring, conductor support 320-6 p. 226
Open wiring, dead end connection 320-6 p.226
Open wiring, maximum voltage 600v 225-10 p.86
Open wiring, porcelain supports 225-12 p.86
Open wiring, running boards 320-14 p.227
Open-circuit voltage 1000 or less 410-73a p. 374
Open-circuit voltage lighting track 410-101c p.377
Open-circuit voltage over 1000v 410-80a p.376
Open-circuit voltage over 300v 410-75 p. 375
Open-conductor, separation 225-14c page 86
Open-conductor, spacing 225-14 page 86
Open-conductor, supports 225-12 page 86
Open-conductors, clearance 225-18 page 87
Open-conductors, final spans 225-19d page 88
Open-conductors, on insulators 225-4 p.84
Open-conductors, on poles spacing 225-14d p.86
Open-conductors, over 600v clear. T.710-33 p.815
Open-conductors, porcelain-knobs 225-12 p.86
Open-conductors, protection 225-20 page 88
Open-conductors, service drops 230-22 p.91
Openings to be closed, boxes 370-18 p.306
Openings to be closed, cabinets 373-4 p.313
Openings to be closed, equip. 110-12a p.39
Operating handles, 50 pounds 710-241 ex.1 p.813
Operating valves, motors T.430-22a ex. p.425
Operation at standstill, motors 430-7c p.419
Opposite polarity spacing T.384-36 page 331
Optical density, signal ciruit 725-71a FPN p.831
Optical fiber cables 770 p.845
Optical fiber cables, conductive 770-4b p.845
Optical fiber cables, nonconductive 770-4a p.845
Optional calculation, add. loads 220-35 p. 81
Optional calculation, dwelling unit 220-30 p. 77
Optional calculation, exist. dwelling 220-31 p.78
Optional calculation, multi-dwelling 220-32 p.79
Optional calculation, restaurant 220-36 p. 81
Optional calculation, school 220-34 page 80
Optional markings 310-11d p.181
Optional standby systems 702-2 FPN p.800
Orange triangle, receptacles 410-56c page 370
Orderly shutdown, elect. system 240-12 p.110
Orderly shutdown, integrated sys. 685-1 p.774
Orderly shutdown, motor 430-44 p.433
Organ, conductor size 650-5a p.736
Organ, overcurrent protection 650-7 p.736
Organ, rectifier grounding 650-4 p.735
Organic coatings boxes raintight 300-6a p. 165
Organic dusts 500-3f p.500
Orientation of conductors 310-4 FPN p.177
Ornamental pools 680-1 FPN p.755
Oscillator blocking, induction heat 665-67 p.743
Oscillator-type units, induct. heat 665-60 p.742
Outbuildings snap sw. disconnect 225-8c ex. p.85
Outdoor antenna conductor T.810-16a p.867
Outdoor installations 225 p.83
Outdoor lampholders, puncture wire 225-24 p.88
Outdoor lighting 225 p.83
Outdoor lighting trees 410-16h p.363

Outdoor portable signs, GFCI 600-10c2 p.692
Outdoor receptacles 210-52e p.65
Outdoor trench, derating fact. Note 8 ex.4 p.196
Outdoor type boxes 300-6a page 165
Outer screw-shell terminal 210-6c2 p.55
Outer sheath properties, cords 400-9 p. 347
Outlet boxes to be covered 410-12 page 361
Outlet boxes, support 50 pounds 410-16a p. 363
Outlet, appliance 210-50c p.63
Outlet, attended 410-57b ex. page 371
Outlet, heat tape 550-8g p.637
Outlets, abandoned 354-7 page 287
Outlets, discontinued 354-7 page 287
Outlets, discontinued 356-7 page 288
Outlets, dwelling 210-52 page 63
Outlets, flush with finish 370-20 p.306
Outlets, laundry 210-52f p.65
Outlets, required 210-50 page 63
Outlets, tree lighting 410-16h p.363
Outline lighting, disconnect 600-6 p.690
Output circuit, induction heating 665-44 p742
Output for bias supplies 665-24 p.740
Outside branch circuits 225 page 83
Outside dimensions floor area 220-3b page 69
Outside feeders 225 page 83
Outside, common neutral 225-7b page 84
Outside, conductor covering 225-4 page 84
Outside, overhead spans 225-6a page 84
Outside, service consider. 2" concrete 230-6 p.91
Outside, service disconnect 230-70a p.99
Outside, wiring on bldgs. 225-10 page 86
Oven receptacles 210-52b2 ex.2 p.64
Ovens, cord connected 422-17a page 383
Ovens, feeder demand T.220-19 page 75
Over 600 v, nonshielded cables 710-4b p.805
Over 600 volts 710 p.804
Over 600v, clearance live parts T.710-33 p.815
Over 600v, electrode boilers 710-70 p.818
Over 600v, minimum cover T.710-4b p.806
Over-pressure limit control 424-74 page 398
Over-temperature limit control 424-73 p.398
Overcurrent device phys. damage 240-24c p. 115
Overcurrent device, vertical pos. 240-33 p. 116
Overcurrent devices, clothes closet 240-24d p.115
Overcurrent devices, readily acc. 240-24a p. 115
Overcurrent devices, wet location 240-32 p. 116
Overcurrent prot, feeder over 600v 240-100 p.119
Overcurrent prot, location from arc 240-41a p.116
Overcurrent prot. fixture wire 240-4 p.109
Overcurrent prot., bathrooms 240-24e p.115
Overcurrent prot., capacitors 460-25c p.488
Overcurrent prot., Class I Div. 2 501-6b4 p. 511
Overcurrent prot., electroplating 669-9 p.749
Overcurrent prot., emerg. systems 700-25 p.796
Overcurrent prot., indust. mach. 670-4b p.750
Overcurrent prot., motors b.c. T.430-152 p.458
Overcurrent prot., second. ties 450-6b p.477
Overcurrent prot., transformer 450-3b p.473
Overcurrent protect., readily acc. 240-10 p. 110
Overcurrent protection, appliances 422-28a p.392
Overcurrent protection, Class 1 725-23 p.822
Overcurrent protection, DC gen. 530-63 p.622
Overcurrent protection, dimmers 520-25 p.602
Overcurrent protection, fire cir. 760-23 p.835

KEY WORD INDEX

Overcurrent protection, hazard 240-3 a p.108
Overcurrent protection, location 240-20 p. 111
Overcurrent protection, organ 650-7 p.736
Overcurrent protection, welders 630-12a p.726
Overcurrent, infrared heating 422-28c p. 386
Overcurrent, nonmotor appl. 422-28e p. 386
Overcurrent, plugging boxes 530-18d p.619
Overcurrent, trip unit 240-20a p.111
Overhead feeders house to garage 225-6a1 p.84
Overhead service drop clear. pool 680-8 p.758
Overhead spans 225-6a p.84
Overheating receptacle 555-4 FPN p.687
Overload protection, motors 430-55 p.437
Overload relay, class 20 or 30 430-34 FPN p.431
Overload sizing, motors 430-32 p.428
Overload units location T.430-37 p.432
Overloads required for 3ø T.430-37 p.432
Oversized knockouts 250-76 ex. page 142
Ozone-resistant insulation 310-6 page 178

-P-

Paddle fans, box support 370-27c page 309
PAF & PTF wire types 760-27c ex. p.863
Paint 110-12c p.40
Paint containers, grounded 516-4f p.563
Paint containers, grounded 516-5d p.564
Paint, gooseneck service head 230-54b p.97
Paint, wallpaper, finished ceilings 424-42 p.395
Paint, white wire 200-7 ex.1 p.50
Pan boxes 370-25b page 309
Panel sets under floor covering 424-99b p.402
Panelboard as service, bonding 384-3c p.325
Panelboard bonding, health care 517-14 p.571
Panelboard, back-fed 384-16f page 329
Panelboard, ceiling clearance 384-8 page 327
Panelboard, clearance working 110-16a p.41
Panelboard, continuous load 384-16c page 329
Panelboard, dead front 384-18 page 329
Panelboard, dedicated space 384-4a page 326
Panelboard, grounding 384-20 page 329
Panelboard, high-leg marking 384-3e page 325
Panelboard, individual prot. 384-16 ex.1 p. 328
Panelboard, lighting & appliance 384-14 p. 328
Panelboard, min. spacing parts T.384-36 p. 331
Panelboard, multiwire branch circuit 210-4 p.53
Panelboard, neutral connections 384-14 p. 328
Panelboard, not as a junction box 373-8 p.315
Panelboard, number of breakers 384-15 p. 328
Panelboard, opposite polarity T.384-36 p.331
Panelboard, overcurrent protection 384-16 p.328
Panelboard, phase arrangement 384-3f p. 326
Panelboard, pool 680-25d p.766
Panelboard, raceway end fitting 384-10 p.327
Panelboard, rating 384-13 page 328
Panelboard, rec. vehicle 551-45 p.655
Panelboard, splices in not allowed 373-8 p.315
Panelboard, sprinkler protect. 384-4 a1 p.326
Panelboard, switches 30a or less 384-16b p.329
Panelboard, used as service 384-3c page 325
Panelboard, wet location 1/4" air 240-32 p.116
Panelboard, wet or damp location 384-17 p.329
Panelboard, working clearance 110-16a p.41
Panelboards 384 page 325

Panelboards, column type 300-3b ex.2 p.160
Panic bars, transformer doors 450-43c p.482
Pans, fixture 410-13 p.361
Pantry, small appliance 210-52b p.63
Paper spacer thickness, gas cable T.325-21 p.233
Paperlined lampholders 410-42b page 368
Parallel power prod. systems 230-2 ex.2 p. 89
Parallel to joists, studs, rafters 334-10f p.246
Parallel, #1/0 conductors 310-4 p.176
Parallel, #20 or larger wire 620-12a1 ex. p.710
Parallel, elevator lighting 620-12a1 ex. p.710
Parallel, existing installations, 310-4 ex. 4 p. 177
Parallel, fuses or breakers 240-8 page 110
Parallel, transformers 450-7 p.477
Parallel, traveling cables 620-12a1 ex. p.710
Paralleling efficiency of rods 250-84 FPN p.146
Park Trailers 552-2 DEF page 68
Park trailer calculations 552-47 page 677
Park trailer factory tests 552-60 page 684
Park trailer labeling 552-44d page 674
Park trailer mast 552-44f page 675
Parking garages, fuel 511-2 p.539
Part-winding motors 430-3 page 414
Parts, broken; bent; cut 110-12c p.40
Patch panel, theater 520-50 p.606
Patching tables, lampholders 530-41 p.621
Patient bed receptacles 517-18b p.572
Patient care area, definition 517-3 p.568
Patient equip. ground. point 517-19c FPN p.573
Patient room toilet and basin 517-21 p. 574
Patient sleeping rooms 517-10 ex. 2 p. 570
Patient vicinity, definition 517-3 p.569
Patient vicinty, exposed surfaces 517-11 p.570
Paving, earth access to box 370-29 p.310
Peak load shaving, emergency 700-5b p.790
Peak radio-frequency output 665-67 p.743
Pediatric location receptacles 517-18c p.572
Pendant boxes, strain relief 370-23g1 p.308
Pendant cond., twisted or cabled 410-27c p.365
Pendant conductors, size 410-27b page 365
Pendant fixtures, Class I location 501-9a3 p.513
Pendant fixtures, closets 410-8c page 361
Pendant lampholders, support 410-27a p. 365
Pendant pushbutton, plating 668-32b3 p.748
Pendant receptacle, double screw 410-38c p.368
Pendants bathtub rim 410-4d page 359
Penetrations are made 300-21 FPN p.171
Peninsular counter top receptacles 210-52c3 p.64
Periodic duty, motors T.430-22a ex. p.425
Permanent barriers, separate box 370-28d p.310
Permanent ladders, access 110-33b p.46
Permanent moisture level, electro. 250-83 p.145
Permanent plaque 230-2b p.90
Permanent plaque, directory 225-8 p.85
Permanently installed feeder 550-5a p.632
Persons who are not qualified 110-31 p.44
Persons, 100 or more assembly 518-1 p.593
Persons, untrained 90-1c p.21
Phase arrangement 430-97b p.445
Phase arrangement panelboard 384-3f p. 326
Phase converter marking 455-4 page 484
Phase converter O.C.P. 455-7 p.484
Phase converter, ampacity 455-6 p.484
Phase converter, capacitors 455-23 p.486

KEY WORD INDEX

Phase converter, motors 455-2 FPN p.483
Phase converter, power interruption 455-22 p.486
Phase converter, start-up 455-21 p.486
Phase converters 455 p.483
Phase converters, static & rotary 455-2 p.484
Phase fault protection over 600v 710-72c p.818
Photovoltaic modules, solar 690-18 FPN p.780
Phychiatric hospital, definition 517-3 p.569
Physical damage, conductors 300-4 page 160
Physical environment de-icing 426-10 (1) p.404
Physical protection requirements T. 300-5 p.163
Piercing a floor, pool cord 680-6 FPN p.757
Piers, receptacles 555-3 p.686
Pig tail required to device 300-13b p.168
Pilot light circuit protection 384-32 page 330
Pilot light, conductors 520-53f ex. p.608
Pilot light, receptacle 520-73 p.613
Pilot light, stage 520-53g p.608
Pipe heating assemblies 422-29 p.387
Pipe organs, wire size 650-5a p.736
Pipeline, definition 427-2 p.409
Piping without valves 500-5b FPN 2 p.501
Pit or depression, garages 511-3b p.539
Pit, aircraft hangar 513-2a p.542
Pit, spray application 516-2a5 p.556
Places of assembly 518-1 p.598
Places of assembly, wiring meth. 518-4 p.599
Plans, feeder diagram 215-5 p.67
Plaque, service directory 230-2b p.90
Plaster 110-12c p.39
Plaster ears, switches 380-10b p.323
Plaster or plasterboard repairs 370-21 p.306
Plaster rings, box 370-16a page 303
Plastic innerduct 770-5 FPN p. 845
Plastic materials 110-11 FPN2 p.39
Plate electrodes, 2 sq.ft. 250-83d p.146
Plate, anode-positive 640-10b p.732
Platforms, access 110-33b page 46
Plenum cable T.725-61 p.829
Plenum, cables 725-71a p.831
Plenums, wiring in 300-22 page 171
Pliable raceway 331-1 p.240
PLTC cables 725-71 p.831
Plug fuse 240-51a page 117
Plug fuse, maximum voltage 240-50 ex. p.116
Plug rating, motor-comp. 440-55b p.468
Plug, 2 wire nonpolarized 422-23 ex. p. 384
Plug, attachment cap 410-56 p.370
Plug, portable X-Ray 517-72c p.597
Plug, wooden not used in mounting 110-13a p.40
Plugging box DEF 530-2 p.617
Plugging boxes, AC system 530-18d p.619
Plugging boxes, cords 530-18d p.619
Plugging boxes, lights 530-18f p.619
Plugging boxes, O.C.P. 530-18d p.619
Plugging boxes, receptacle 530-18 p.618
Plugmold raceways 352-1 page 280
Point of attachment 545-6 ex. p.627
Point of connection DC 250-22 page 126
Point of entrance 225-8b p.85
Point of entrance definition 800-2 p.852
Polarity of appliances 422-24 p.384
Polarity of connections 200-11 page 52
Polarity, thru bushed holes 230-54e page 98

Polarization of fixtures 410-23 page 364
Polarized attachment plugs 200-10a & b p. 51
Polarized or grounding type plug 422-23 p. 384
Polarized, adapter 410-58b3 page 372
Polarized, uniquely 604-6b page 696
Pole base 410-15b1 page 362
Poles, climbing space 225-14d p.86
Poles, comm. cond spacing 800-10a1 p.853
Poles, conductor separation 225-14d p.86
Poles, handhole 410-15b1 p.362
Polymeric braces, wet locations 370-23b2 p. 307
Polyphase, transformer 450-3 p.471
Pool, bonding forming shells 680-22a2 p.763
Pool, bonding metal parts 680-22a6 p.763
Pool, common bonding grid 680-22b p.764
Pool, conduit entries 680-21d p.763
Pool, cord length 680-7 p.758
Pool, cord strain relief 680-21e p.763
Pool, diving structures 680-22a6 p.763
Pool, double insulated pump 680-28 page 767
Pool, dry niche fixture 680-20c p.762
Pool, electrical covers 680-26 p.767
Pool, encapsulated termination 680-20b1 p.761
Pool, equipment room drainage 680-11 p.760
Pool, equipment to be grounded 680-24 p.765
Pool, fixture lens 18" 680-20a3 p.761
Pool, fixture maximum voltage 680-20a2 p.761
Pool, forming shell 680-20b1 p.761
Pool, forming shell un-water sound 680-23b p.764
Pool, GFCI receptacles location 680-6a3 p.657
Pool, grid ground 680-22b p.764
Pool, grounding 680-24 p.765
Pool, grounding terminals 680-21d p.763
Pool, junction boxes 680-21a4 p.762
Pool, light fixture lens 680-20a3 p.761
Pool, light fixtures location 680-6b p.658
Pool, lighting ENT 680-25b1 ex. 2 page 765
Pool, lighting grd. conductor 680-25b1 p.765
Pool, maximum fixture voltage 680-20a2 p.761
Pool, normal water level 680-20a3 p.761
Pool, observation stands 680-22a6 p.763
Pool, overhead conductors 680-8 p.758
Pool, panelboard grounding 680-25d p.766
Pool, potting compound 680-20b1 p.761
Pool, pump recpt. 680-6a1 ex. p.757
Pool, receptacles from wall 680-6a1 p.757
Pool, recirculating pump recpt. 680-6a1 ex. p.757
Pool, steel tie wires 680-22a6.x.1 p.763
Pool, storable definition 680-4 p.756
Pool, storable GFCI 680-31 p.767
Pool, switching devices 680-6c p.758
Pool, therapeutic 680-60 p.772
Pool, transformer 680-5a p.757
Pool, underwater audio 680-23 p.764
Pool, underwater lights 680-20a1 p.760
Pool, underwater speakers 680-23a p.764
Pool, wading definition 680-4 p.756
Pool, water heaters 680-9 p.760
Pool, wet-niche fixture 680-20b p.761
Pools, swimming 680 p.755
Porcelain covers, boxes 370-41 page 311
Porcelain fixture, live parts 410-46 page 369
Porcelain knobs, open conductors 225-12 p.86
Porcelain lampholders 410-72 page 374

KEY WORD INDEX

Portable appliance, 80% of circuit 210-23a p.61
Portable appliance, definition 550-2 p.631
Portable cables over 600v 400-30 page 349
Portable generators 250-6a page 124
Portable lamps 410-42 page 368
Portable lamps, cords 400-7a page 347
Portable lamps, spray area 516-3d p.562
Portable motor 1/3 hp 430-81c p.441
Portable outdoor signs GFCI 600-10c2 p.692
Portable power cable T.400-4 page 336
Portable power cable T.400-4 page 341
Portable signs, 600-10 p.692
Portable switchboards, neutral 520-53o p.610
Portable tools wet location 250-45d p. 134
Positive mechanical ventilation 516-2e p.561
Positive-pressure ventilat. 500-2a3 FPN 1 p.494
Potential coils switchboard 384-32 page 330
Potential difference 517-11 p.570
Potential differences 250-86 FPN2 page 146
Potential differences 517-19c FPN p.573
Potential transformer protection 384-32 p. 330
Potential transformer, fuses 450-3c p.474
Potting compound, pool 680-20b1 p.761
Pound-inches, torque motor term. 430-9c p.420
Powder coating 516-6 p.564
Power lines, switchboard 520-53o p.610
Power and Control tray cable 340-4 page 255
Power conductors, spacing 225-14d page 86
Power conversion equip. 430-2 page 414
Power conversion equip. 430-22a ex. 3 p. 425
Poer factor correction capacitors 430-2FPN p. 414
Power fuses in parallel 710-21b1 p.809
Power interruption, phase converter 455-22 p.486
Power limited exposed cable 760-30a2 p.837
Power loss hazard 240-3 a page 108
Power service drop conductors 230-28 FPN p.93
Power supply cord, mobile homes 550-5b p.633
Power-driven machine 670-2 p.749
Power-limited circuits 725 p.819
Power-limited tray cable T.725-61 p.829
Predetermined cells, raceway 356-1 p.288
Premises, laundry Example 4a p.903
Preservation of the safety 310-15a4 p.190
Pressure connectors 110-14c3 p.41
Pressure plates 424-59 FPN page 396
Pressure-relief vent 450-25 p.480
Pressurized sulfur hexafluoride gas 325-21 p.233
Prevent energizing of the machine 501-8a p.512
Preventing pull on a cord 400-10 FPN p. 347
Primary leads, ballast 300-3c2 ex.2 p.160
Prime mover 700-12b1 p.793
Principal exit doors 645-10 p.734
Products of combustion 300-21 page 171
Projection ports 540-10 p.624
Projectors, incandescent-type 540-13 p.625
Prolonged submersion, motor T. 430-91 page 444
Prongs, blades, or pins attach. plug 410-56f p.370
Propagation of flame 501-5d ex. p.509
Propagation of flames 501-5 FPN2 p.506
Propane 500-3a4 p.496
Proper maintenance 90-1b p. 21
Propylene oxide 500-3a2 p.496
Proscenium 520-2 DEF page 601
Proscenium lights, stage 520-41 p.604

Protection substantially equivalent 373-4 p.313
Protective covers, flat cable 363-18 p.296
Protector grounding cond. 800-30b FPN p.856
Protector location 800-30b FPN p.856
Psychiatric security rooms 517-18b ex.2 p.572
Public address, sound 640-1 p.730
Pull and junction boxes 370-28 page 309
Pull point 300-15b page 168
Pull-type canopy switches 410-38c page 368
Pump, dispensing disconnect 514-5 p.549
Pumps, canned 501-5f3 p.510
Pumps, fountain 680-51b p.771
Pumps, submersible 501-11 p.515
PVC conduit, expansion joints 347-9 page 269
PVC conduit, exposed 347-2f page 268
PVC conduit, joints 347-6 page 269
PVC conduit, number of bends 347-14 p. 270
PVC conduit, number of conductors 347-11 p.269
PVC conduit, splices & taps 347-16 page 270
PVC conduit, support T.347-8 page 249
PVC conduit, support of fixtures 347-3b p.248
PVC conduit, trimming 347-5 page 269
PVC rigid conduit 347-1 page 267
PVC thermal expansion, Table 10 p.890

-Q-

Qualified personnel 520-53p p.610
Qualified personnel 690-13 FPN p.779
Qualified persons, isolating switch 230-204c p. 105
Qualified testing laboratories 300-21 FPN p.171

-R-

Raceway supported enclosure 370-23d p. 307
Raceway to open wiring 300-16 page 169
Raceway transitions, direct burial 300-5j p.165
Raceway, pliable 331-1 p.240
Raceways arranged to drain, bldgs. 225-22 p.88
Raceways arranged to drain, service 230-53 p.97
Raceways, complete system 300-18 page 169
Raceways, continuity 300-10 page 166
Raceways, different temperatures 300-7 p.166
Raceways, end fitting into panel 384-10 p.327
Raceways, flat-top 354-3a page 286
Raceways, half-round 354-3a page 286
Raceways, induced currents 300-20a page 171
Raceways, installed complete 300-18 p.169
Raceways, listed fixtures 410-31 ex.1 p. 367
Raceways, means of support 300-11b page 167
Raceways, on exterior surfaces 225-22 page 88
Raceways, other systems 300-8 page 166
Raceways, sealing 300-7a page 166
Raceways, securing 300-11a page 167
Raceways, spacing between Note 8b page 196
Raceways, strut type channel 352-40 page 283
Raceways, trench-type 354-3c page 286
Racked, conductors 110-12b p.39
Racks, open-conductors 225-12 p.86
Radiant heating panels 424-90 page 400
Radiation or conduction, spray area 516-3c p.561
Radii of bends, flexing use 349-20 page 274
Radio & Television antennas 225-19b p.88
Radio & Television equipment 810 p.865

KEY WORD INDEX

Radio-frequency converters 665-67 p.743
Radiographic, X-ray 660-23a p.738
Radius of bends, armored cable 333-8 p.243
Radius of bends, conductors 110-3a3 p.38
Radius of bends, conduit T.346-10 p.265
Radius of bends, flex. metal T.349-20a p.274
Radius of bends, gas cable T.325-11 p.232
Radius of bends, hi-voltage cable 300-34 p.173
Radius of bends, lead-covered 300-34 p.173
Radius of bends, metal clad cable 334-11 p.246
Radius of bends, MI cable 330-13 p.239
Radius of bends, nonshielded cable 300-34 p.173
Radius of bends, romex 336-16 p.241
Radius of bends, shielded cable 300-34 p.173
Radius of the curve 346-10 page 265
Rail serving as conductor 610-21f4 p.703
Railway conductors 110-19 page 44
Rain, snow, sleet motor enclosure T.430-91 p.444
Rainproof 300-6a page 165
Raintight 300-6a page 165
Raintight service head 230-54a page 97
Raised covers single screw 410-56i p.371
Raised floor, computer 645-5d p.733
Range, 1 ø on 3 ø system 220-19 p.74
Range, disconnect 422-22b p.384
Range, dryer cable T.400-4 page 340
Range, equip. grd. conductor-neutral 400-5 p.346
Range, feeder demand T.220-19 p.75
Range, grounded conductor 250-60 p.138
Range, neutral 70% 220-22 p.76
Range, taps 210-19b ex.1 page 59
Rated input to power conversion 430-2 p.414
Rated over one ampere 422-31 p. 387
Rated-load current 440-6a ex.1 p.460
Rated-load current A/C 440-2 p.458
Ratings in volts & amps 424-29 page 392
Ratings in volts & watts 424-29 page 392
Re-bar electrodes 250-81c2 p.145
Reactor, space separation 470-3 p.489
Reactors, metallic enclosures 470-18e p.490
Reactors, temperature rise 470-18e p.490
Readily accessible disconnect, A/C 440-14 p. 463
Readily accessible location 225-8 p.85
Readily accessible, motor switch 430-107 p.447
Readily accessible, O.C.P. 240-24a p.115
Readily accessible, switches 380-8a page 322
Readily distinguishable solid color 400-22a p.348
Readily identified, emerg. systems 700-9a p.791
Readily ignitible residues 516-6a ex. p.564
Reamed, rigid conduit 346-7a page 264
Reasonable efficiency,volt. drop 230-31 FPN p.94
Receiving sta. outdoor antenna T.810-16a p.867
Receptacle loads, demand factor T.220-13 p.73
Receptacle outlet cord pendant 210-50a p.63
Receptacle rating, motor-comp. 440-55b p.468
Receptacle, 15a on 20a circuit T.210-21b2 p.60
Receptacle, adjacent to basin 210-52d p.65
Receptacle, basement 210-52g page 65
Receptacle, bathroom 210-52d page 65
Receptacle, bathroom/shower space 410-57c p.371
Receptacle, bonding at box 250-74 p.140
Receptacle, circuit rating T.210-21b3 page 60
Receptacle, counter top space 210-52c p.64
Receptacle, floor 210-52a page 63

pg - 31

Receptacle, garage 210-52g page 65
Receptacle, hallway 210-52h page 65
Receptacle, heat-AC 210-63 page 65
Receptacle, individual b.c. 210-21b1 p.60
Receptacle, laundry 210-52f page 65
Receptacle, malfunctioning 555-4 FPN p.687
Receptacle, maximum load T.210-21b2 p. 60
Receptacle, nongrounding type 210-7d p.56
Receptacle, outdoors 210-52e page 65
Receptacle, overheating 555-4 FPN p.687
Receptacle, pilot light 520-73 p.613
Receptacle, rating T.210-21b3 p.60
Receptacle, replacement 210-7d p.56
Receptacle, rooftop 210-63 p.65
Receptacle, self-closing cover 410-57b p.371
Receptacle, single 210-21b1 page 59
Receptacle, T-slot 410-56h ex. page 371
Receptacles 210-7 page 55
Receptacles, AC or DC 210-7f page 57
Receptacles, anesthetizing loc. 517-61a5 p.591
Receptacles, attachment plugs 410-56f p 370
Receptacles, cigarette lighter 551-10h p.650
Receptacles, cleat-type 410-3 ex. p.358
Receptacles, clock 210-52b2 ex.1 p.64
Receptacles, CO/ALR 410-56b page 370
Receptacles, construction sites 305-4d p.174
Receptacles, counter top space 210-52c p.64
Receptacles, critical care area 517-19b p.573
Receptacles, damp location 410-57a page 371
Receptacles, demand factor 10kva 220-13 p.73
Receptacles, demand shore power 555-5 p.687
Receptacles, different frequencies 210-7f p. 57
Receptacles, dressing rooms 520-73 p.613
Receptacles, exemption GFCI 210-8a ex.2 p.57
Receptacles, face up position 210-52c5 p.64
Receptacles, face-up 550-8f2 p.637
Receptacles, faceplates 410-56d page 370
Receptacles, floor protection 410-57d page 371
Receptacles, frequencies 210-7f p.57
Receptacles, gas-fired ranges 210-52b2 ex.2 p.64
Receptacles, general care area 517-18b p.572
Receptacles, grounding poles 410-58a page 372
Receptacles, grounding type 410-29b page 366
Receptacles, grounding type 210-7a p.55
Receptacles, guest rooms 210-60 page 65
Receptacles, height dwelling 210-52a p.62
Receptacles, hospital grade 517-18b p.572
Receptacles, hotels/motels 210-60 p.65
Receptacles, ignition system 210-52b2 ex.2 p.64
Receptacles, island counter top 210-52c2 p.64
Receptacles, isolated ground 410-56c page 370
Receptacles, kitchen counter 210-52c page 64
Receptacles, location 5 1/2' 210-52a p.63
Receptacles, locking & grd. type 555-3 p.686
Receptacles, marinas 555-3 p.686
Receptacles, mobile homes 550-8 p.636
Receptacles, nongrounding type 210-7d p.56
Receptacles, office furnishings 605-5c p.697
Receptacles, orange triangle 410-56c page 370
Receptacles, outdoors dwelling 210-52e p.65
Receptacles, ovens 210-52b2 ex.2 p.64
Receptacles, patient bed 517-18b p.572
Receptacles, pediatric locations 517-18c p.572
Receptacles, peninsular counter top 210-52c3 p.64

KEY WORD INDEX

Receptacles, plugging boxes 530-14 p.618
Receptacles, pool 680-6a1 p.757
Receptacles, pool GFCI 680-6a3 p.757
Receptacles, portable cords 410-56a page 370
Receptacles, protect from weather 410-57a p.371
Receptacles, raised covers 410-56i p.371
Receptacles, required dwelling 210-52 p.63
Receptacles, shore power 555-3 p.686
Receptacles, show window 12' 210-62 p.65
Receptacles, showcases 410-29b page 366
Receptacles, spa or hot tub 680-41a p.768
Receptacles, stage 520-45 p.606
Receptacles, tamper resistant 517-18c p.572
Receptacles, water accumulation 410-57f p. 371
Receptacles, wet location 410-57b page 371
Receptacles, rec. vehicle 551-41 p.653
Recepts., baseboard heater 210-52a FPN p. 63
Recepts., baseboard heaters 424-9 FPN p. 388
Recepts., flush mount faceplate 410-57e p. 371
Recepts., weatherproof faceplate 410-57e p.371
Recessed fluorescent fixture 410-8b2 page 360
Recessed incandescent fixture 410-8b1 p. 360
Recessed incandescent fixtures 410-8d3 p.361
Reciprocators, spray application 516-4b p.562
Recording equipment 640-1 p.730
Recreational vehicles, appl. 551-2 FPN p.646
Recreational vehicles, auto trans. 551-20e p.651
Recreational vehicles, frame 551-2 p.646
Recreational vehicles, GFCI 551-41c p.653
Recreational vehicles, panelboard 551-45 p.655
Recreational vehicles, parks 551-2 Def. p.647
Rectifier bridge, motors 430-22a ex.2a p.424
Rectifier filter components 665-24 p.740
Rectifier voltage, motors 430-18 p.424
Rectifier, auxiliary 665-24 p.740
Rectifier, ground 250-3 ex.4 p.123
Rectifier, grounding organ 650-4 p.735
Rectifier, transformer-type organ 650-3 p.735
Reel 347-1 p.267
Reels 343-1 p.258
Reflecting systems, solar 690-52 FPN p.782
Reflection pools 680-1 FPN p.755
Refrigerating equipment 440-1 p.458
Refrigerators, rec. vehicle 551-2 FPN p.647
Regenerated energy, elevators 620-91a p.721
Relays, motor overload 430-40 p.432
Release-type adhesive, FCC cable 328-10 p.235
Remote-control circuits 725 p.819
Remote-control conductors 430-72b p.439
Remote panel, grounding cond. 680-25d p.766
Removal of a drawer, ranges 422-22b p.384
Removed from the raceway, outlets 354-7 p.287
Removed from the raceway, outlets 356-7 p.288
Repairing drywall no gaps 1/8" 370-21 p.306
Replaceable in the field 422-31 p. 387
Replaceable in the field 424-29 page 392
Reproduction, sound 640-1 p.730
Requirements, physical protection T. 300-5 p.163
Requisite mechanical strength 410-36 p. 367
Residual voltage, capacitors 460-6a page 487
Residues 516-3b p.561
Resist.-type heating appliances 422-28f p.386
Resistance elements, heat 424-22b page 384
Resistance of conductors, AC Table 9 p.889

Resistance of conductors, DC Table 8 p.888
Resistance of made electrodes 250-84 p.146
Resistance welder 630-31 p.728
Resistance, AC Table 9 p.889
Resistance, DC Table 8 p.888
Resistance, ground rod 25 ohm 250-84 p.146
Resistance-type boilers 424-70 page 397
Resistance temperature devices 504-2FPN p. 532
Resistant to crushing, PVC 347-1 page 267
Resistant to distortion, PVC 347-1 page 267
Resistor duty, motors T.430-23c p.426
Resistor, bleeder 665-24 p.740
Resistors and Reactors 470 p.489
Resistors, space separation 470-3 p.489
Resonance, motors 430-2 FPN page 414
Restrict the sizing 110-14c FPN p.41
Rheostats, motor-starting 430-82c p.441
Right angles to the cells 358-5 page 290
Rigid conduit, bends 346-10 page 265
Rigid conduit, buried T.300-5 p.163
Rigid conduit, durably identified 346-15c p. 267
Rigid conduit, reamed 346-7 p.264
Rigid conduit, standard length 346-15a p. 267
Rigid conduit, supports T.346-12 page 267
Rigid metal conduit corrosion 346-1c p.263
Rigid metal conduit stem 501-9a3 p.513
Rigid metal conduit, size 346-5 page 264
Rigid nonmet. conduit, not permit.347-3b ex. p.268
Rigid nonmetallic conduit 347-1 page 267
Rigid nonmetallic conduit, install. 347-5 p.268
Rigid structural system, cable tray 318-2 p.216
Ring, box extension 370-22 p.307
Riser cable, marking T.725-71 p.832
Risk of ignition 501-8b FPN2 p.513
Rivets, screws, bolts fasten lights 410-16c p.363
RMS open-circuit voltage 410-101c p.377
Road show connection, theaters 520-50 p.606
Robot, arm 516-4 p. 562
Robotic, devices 516-4 p. 562
Robotic, programing 516-4 p. 562
Rock bottom, ground rod 250-83c3 page 146
Rod and pipe electrodes 250-83c page 146
Rolling stock 90-2b1 page 21
Romex, ampacity 336-30b page 251
Romex, attic joists 336-18 p.250
Romex, bends in cable 336-16 p.250
Romex, fished in air voids 336-4a p.248
Romex, in cement blocks 336-4a p.248
Romex, in raceway grounding 250-33 ex.1 p.131
Romex, knockout opening 370-17c page 306
Romex, protection from damage 336-6b p. 249
Romex, sizes #14 - #2 336-30b p.251
Romex, supports 336-18 page 250
Romex, uses 336-4 page 248
Roof required over equipment 110-11 p.39
Roof/ceiling assembly 300-11a p.167
Roofed open porches 410-4a page 358
Rooftop outlet 210-63 p.65
Rooftop receptacle 210-63 p.65
Room air conditioners, cords 440-64 p.469
Room air conditioners, disconnect 440-62a2 p.468
Rooms for bulk chemicals 300-6c FPN p.166
Rooms for casings 300-6c FPN page 166
Rooms for fertilizer 300-6c FPN page 166

KEY WORD INDEX pg - 33

Rooms for hides 300-6c FPN page 166
Rooms for salt 300-6c FPN page 166
Root-mean-square current 620-13 FPN p.710
Root-mean-square, voltage DEF 100 p.35
Ropes, chains or sticks busways 364-12 p.298
Rostrums 513-6a p.543
Rotary converter leads 640-6b p.731
Rotary, phase converter 455-2 page 484
Rotary, phase converter 455-2 DEF p.483
Round boxes 370-2 page 302
Rubber mats, motors 430-133 p.452
Rubber mats, switchboards 250-123c p.154
Rubber-filled cords T.400-4 note 3 p. 332
Running boards 300-5f p. 162
Running boards 320-14 p. 227
Running boards, AC cable 333-11 ex.1 p.244
Running overload prot. shunted 430-35a &b p.431
Running threads conduit 346-9b p.265

-S-

S loops direct burial 300-5j FPN p.165
S type fuse 240-53 p.117
Safeguarding of persons & property 90-1a p. 21
Safety combustion controls 430-32c3 p.430
Scatter box DEF 530-2 p.616
Scatter of cell parts, batteries 480-9b p.492
Scene docks, "back stage lamps" 520-47 p.606
Scenery, bracket wiring 520-63a p.611
Schedule 80 PVC conduit 300-5d p.162
Schedule 80 PVC conduit 347-1 page 267
Screw shell, fuseholder 240-50e p.117
Screw shell, lampholder 410-23 page 364
Screw shells 200-10c page 51
Screw-shell lampholder 210-6c4 p.55
Scuttle hole 333-12a page 244
SE cable, interior wiring 338-3a p.252
SE cable, supply appliances 338-3c page 252
SE cable, temperature limitations 338-3c p.252
Seal unused knock outs 370-18 p.306
Seal unused knock outs 373-4 p.313
Sealing 2" conduit Class I Div.1 501-5a2 p.506
Sealing against dampness 328-11 page 235
Sealing compound 501-5 p.506
Sealing compound, thickness 501-5c3 p.508
Sealing of strands 501-5 FPN2 p.510
Sealing raceways 300-7a page 166
Seals, Class I, Div.2 501-5 p.507
Seals, Class I, Div.I 501-5a1 p.506
Seashore areas 300-6c FPN page 166
Second ground fault 517-19f FPN p.574
Secondary circuits, wound-rotor 430-32d p.430
Secondary ties, O.C.P. 450-6b p.477
Selected receptacles, definition 517-3 p.569
Self-excited 705-40 FPN2 p.804
Self-propelled machinery 90-2b2 p.21
Self-propelled vehicles 511-1 p.539
Self-restoring ground. contact 250-59 ex. p.138
Self-service stations 514-5b p.549
Sense motor rotation 430-35 ex.2a p.431
Sensors, occupancy 210-70a ex.3 p.61
Separable locking-type connectors 410-29 p.365
Separate building, grounding 250-24c p.128
Separate compartments, raceways 352-26 p. 282

Separate services same electrode 250-54 p.136
Separately bushed openings 230-54e p.98
Separately derived system DEF 100 p.33
Separately derived system grd. elect 250-80c p.143
Separately derived systems 250-5d page 124
Separately derived systems 60v 530-70 p. 622
Separation between showcases 410-29c p.366
Sequential delayed automatic 517-34a ex. p.581
Serious degradation 310-10 FPN page 179
Service cables support 230-51a p.96
Service cond., manuf. build. 545-6 ex. p.627
Service conductors, unspliced 230-46 page 95
Service equip., marking 550-23g p.645
Service equipment marking 230-66 p.99
Service factor, motors 430-32a1 p.428
Service fuses 230-82 ex.2 p.101
Service head, attachment point 230-54c ex. p.97
Service head, raintight 230-54a page 97
Service lateral, definition 100 p.34
Service mast 230-28 FPN p.93
Service point DEF 100 p.34
Service size minimum 60 amp 230-42b3 p. 95
Service, floating dock 555-10 p.688
Service, gooseneck 230-54b page 97
Service-drop attachment 24" 230-54c ex. p.97
Service-drop conductors 230-21 page 91
Service-ent. cables, protection 230-50a p.96
Service-ent. cond., manuf. bldgs. 545-6ex p. 627
Service-entrance cable definition 338-1 p.251
Service-entrance cables, supports 230-51a p.96
Service-entrance cond. 600v O.C.P. 230-208 p.106
Service-lateral cond., minimum 230-31b pg.94
Service-lateral voltage drop 230-31 FPN p.94
Services over 15,000 volts 230-12 p.106
Services, alum. neutral undergrd. 230-30ex.d p.93
Services, clearances from bldgs. 230-9 page 91
Services, clearances from ground 230-24b p.92
Services, connections to terminals 230-81 p. 101
Services, disconnect 230-70 p.99
Services, disconnect GFCI 230-95 FPN3 p.104
Services, disconnect max. number 230-71a p.99
Services, disconnect neutral 230-75 page 100
Services, disconnect simultaneously 230-74 p.100
Services, drip loops 230-54f page 98
Services, drop point of attachment 230-26 p.93
Services, electrical continuity 250-72 p.140
Services, emergency supply 230-94 ex.3 p.103
Services, fire pump & equip. 230-94 ex.4 p.103
Services, floating buildings 553-5 p.685
Services, GFCI solid-wye 1000a 230-95 p.104
Services, grounding conductor size T. 250-91 p.151
Services, group disconnects 230-72 page 100
Services, handle ties 230-71b p.99
Services, head point of attach. 230-54c ex. p.97
Services, heights vertical 230-24b p.92
Services, high-voltage 230-200 page 104
Services, indicate open or closed 230-77 p. 100
Services, insulation or covered 230-22 p. 91
Services, load management 230-94 ex.3 p.103
Services, locked O.C.P. 230-92 page 103
Services, lugs not solder 230-81 p.101
Services, masts as supports 230-28 page 93
Services, means of attachment 230-27 page 93
Services, mechanical strength 230-23a p. 92

KEY WORD INDEX

Services, minimum conductor size 230-202a p.105
Services, minimum size 60 amp 230-42b3 p.95
Services, mobile homes 550-21 p.643
Services, neutral size minimum 230-42c p. 95
Services, neutral size overhead 230-23c p. 92
Services, number allowed 230-2 p.89
Services, other conductors 230-7 page 91
Services, outside a bldg. 230-6 page 91
Services, over 600v minimum size 230-202a p.105
Services, over top of window 230-9 ex. page 91
Services, pressure connectors 230-81 page 101
Services, rating of disconnect 230-79 p.101
Services, shore power 555-5 p.687
Services, short-circuit current avail. 230-65 p.99
Services, simultaneously disconnect 230-74 p.100
Services, six operations of hand 230-71b p.99
Services, size and rating 230-42b page 95
Services, spliced 230-46 ex.3 p.95
Services, surge arresters 230-82 ex.4 page 101
Services, taps to main 230-46 ex.2 p.96
Services, top of window 230-9 ex. p.91
Services, undergrd. wiring meth 230-46ex.3 p.96
Services, underground lateral 230-30 page 93
Services, vertical clearances 230-24 page 92
Set-screws, double for canopy 410-38c p.368
Severe corrosive influences IMC 345-3b p.261
Severe deterioration 110-11 FPN2 p.39
Severe over-voltage 705-40 FPN2 p.804
Sewerage disposal 701-2 FPN p.796
Shaded-pole motor 430-6a ex.2 page 417
Shaft seals compressor 440-2 p.459
Sharply angular substance, backfill 300-5f p.162
Shaver, double insulated 422-23 ex. p. 384
Sheet-metal auxiliary gutters 374-9e1 p. 320
Sheet-metal screws 250-113 p.152
Sheet-metal troughs, wireways 362-1 p.291
Shield system, FCC cable 328-14 p. 236
Shielded, cables over 2000v 400-31b page 350
Shielded, conductors 300-34 page 173
Shielded, fuses or breakers 240-41a page 116
Shielding, conductors 310-6 page 178
Shields shall be grounded 504-50c p.534
Shields, cables over 2000v 400-31b page 350
Shock hazard during relamp 680-20a1 p.760
Shore power, receptacles 555-3 p.686
Short circuits, free from 110-7 page 39
Short radius conduit bodies 370-5 p.302
Short sections of raceway 300-12 ex. p.167
Short-circuit current available 230-65 p.99
Short-time duty, motors T.430-22a ex. p.425
Show window lighting, feeder 200va 220-12 p. 73
Show window receptacles 12' 210-62 p.65
Showcases, cord support 410-29c page 366
Showcases, female fitting 410-29c page 366
Showcases, free lead at the end 410-29c p.366
Showcases, receptacles 410-29b page 366
Shower space receptacle 410-57c p.371
Shunting means, cell line 668-14 p.746
Shunting, motors 430-35 p.431
Shunting, service 230-94 ex.2 p.103
SI units B. examples p.897
Side gutters, cabinets 373-11d Ex. p. 317
Side of a building 225-19 ex.4 p.87
Side wiring spaces, cabinets 373-11d ex. p. 317

Sidewalks, excavating 370-29 p.310
Sidewalks, moving 620-1 p.706
Sign & outline lighting load 220-3c6 p.69
Sign, aluminum thickness 600-8c p.691
Sign, GFCI location 600-10c2 p.693
Sign-outline lighting enclosure 300-3c2 ex.2 p.160
Signal circuits, traveling cable 620-12a2 p.710
Signal fixtures 620-2 p.708
Signal, heated appliance 422-12 page 382
Signaling circuits 725 p.819
Signals, derangement 700-7a p.790
Significant capacitance 705-40 FPN2 p.804
Sign boxes, terminals and conductors 600-8a p.691
Signs 600 p.689
Signs, ballasts 600-21 p.692
Signs, accessible to pedestrians 600-5a p.690
Signs, attic location ballasts 600-21e page 693
Signs, branch circuit rating 600-5b1 p.690
Signs, breaker in sight of 600-6a p.690
Signs, clearance 225-19b p.88
Signs, computed load 600-5b3 p.690
Signs, cutouts-flashers 600-6b p.691
Signs, discharge lighting trans. 600-6b1 p.690
Signs, disconnect 600-6 p.690
Signs, drain holes 600-9d1 p.692
Signs, electric discharge lighting 600-2 DEF p.689
Signs, electromech. control 600-6a p.690
Signs, elevation above vehicles 600-9a p.691
Signs, enclosure wood 600-9c p.692
Signs, exposed to weather 600-9d p.692
Signs, flashers-cutouts 600-6b p.691
Signs, GFCI 600-10a2 page 692
Signs, listed 600-3 p.689
Signs, location 600-9 p.691
Signs, marking 600-4 p.689
Signs, material 600-8b p.691
Signs, metal poles for conductors 600-5c3 p.690
Signs, neon tubing 600-2 DEF p. 689
Signs, neon tubing 600-41 p.694
Signs, outdoor portable GFCI 600-10c2 p.692
Signs, outlet required 600-5a p.690
Signs, over 600v transformers 600-23 p.693
Signs, skeleton tubing 600-2 DEF p. 689
Signs, skeleton-type 600-30 p.693
Signs, supply leads enclosed 600-8 p.691
Signs, switches on doors 600-6c p.691
Signs, tube support 600-41b p.694
Signs, tubing 600-41 p.694
Signs, wood decoration 600-9c p.692
Signs, working space 600-21d page 692
Sill height, transformer 450-43b p.482
Simple apparatus 504-2 page 532
Simple reactance ballasts 410-73e ex.1 p.374
Simulating lighting, theaters 520-66 p.612
Simultaneously, disconnect 210-4b page 54
Simultaneously, disconnect 230-74 page 100
Simultaneously, disconnect lamph. 410-48 p.369
Single conductor, ferromagnetic 427-47 p. 413
Single conductors, installation 300-3a p. 159
Single enclosure, limit circuits 90-8b p.23
Single machine, A/C system 440-8 p.461
Single receptacle 210-8a2 ex.2 p.57
Single screw raised covers 410-56i p.371
Single-pole separable connectors 520-53k p.609

KEY WORD INDEX pg - 35

Single-pole separable connectors 530-22 p.620
Sink, wet bar 210-8a7 p.57
Skeleton tubing 600-2 DEF page 689
Skeleton-type, signs 600-30 p.693
Skin effect heating 426-40 page 407
Skin effect heating definition 426-2 p.404
Slash rating circuit breaker 240-85 p.119
Sliding panels receptacle spacing 210-52a p.63
Small appliance branch circuits 220-4b p. 71
Small appliance, feeder load 1500va 220-16a p.73
Small appliances 20 amp 210-52b p.63
Smoke characteristics, tray cable 340-6 p.255
Smoke control 517-34a4 page 581
Smoke generation, ENT 331-1 page 240
Smoke removal systems 701-2 FPN p.796
Smoke ventilator control 520-49 p. 606
Smoothing iron stand 422-11 page 381
Smoothing irons 50 watts 422-8a p.380
Smoothing irons identified temp.limit 422-13 p382
Snap switch as disconnect 225-8c ex. p.85
Snap switches ganged 300v 380-8b p.322
Snap switches rated 347 volts 380-14d p.324
Snow-melting equipment 426 page 403
Soft-drawn copper, antenna 810-11 ex. p.866
Solar cell, definition 690-2 p.776
Solar photovoltaic systems 690 p.775
Solar system, ampacity 690-8a p.778
Solar system, O.C.P. 690-8a p.778
Solar, neutral ampacity 690-62 p.782
Solar, storage batteries 690-71 p.784
Solar, unbalanced interconnect. 690-63 p.783
Solder, depend upon 230-81 page 101
Soldered splices joined mechanical 110-14b p.40
Solely by enamel, boxes 370-40a FPN p. 311
Solely by enamel, EMT protection 348-1 p.270
Solid dielectric insulated conductors 310-6 p.178
Solid-state devices & tubes 665-28 p.741
Solidly grounded 250-1 FPN1 page 120
Solidly grounded neutral 250-152 p.155
Solidy grounded neutral 230-95 p.104
Sound recording, cord size 640-7 p.731
Sound-recording, wireways-gut. 640-4 ex.a p.730
Sound-recording 640-1 p.730
Sources of heat, discharge lamps 410-54a p.370
Spa-hot tub, water heaters 680-41 h p.770
Space heating, branch circuit 424-3 page 388
Space heating, feeder 220-15 page 73
Space separation, resistors 470-3 p.489
Spacing bare metal parts T.384-36 page 331
Spacing between conduits note 8b p.196
Spacing bus bars motors T.430-97 p.445
Spacing for conductor supports T.300-19a p.170
Spark gap, surge arrester 280-24b p.158
Sparks, open motors 430-14b p.423
Spas & hot tubs 680-40 p.768
Spas, GFCI 680-42 p.770
Speakers, underwater pool 680-23a p.764
Special effects, stage 520-66 p.612
Special perm., lead covered 640-3 ex.1&2 p.730
Special permission from inspector 90-2c p. 22
Special permission, conductor size 358-10 p.290
Specific inductive capacity 310-6 ex.c p.178
Specification, design 90-1c p.21
Speech-input systems, sound 640-1 p.730

Spider (cable splicing block) 530-2 DEF p.617
Spiders 530-15d p.618
Splice or tap loop wiring 354-6 p.287
Splices & taps, antennas 810-14 p.866
Splices & taps, boxes 347-16 p. 270
Splices & taps, buried cables 300-5e p. 162
Splices & taps, conduit bodies 370-16c p.305
Splices & taps, cords 400-9 p.347
Splices & taps, gutters 374-8 p.319
Splices & taps, headers 356-6 p. 288
Splices & taps, raceway 300-13a p.167
Splices & taps, services 230-46 p.95
Splices & taps, wireways 362-7 p. 292
Splices equivalent to 110-14b page 40
Splices, embedded cables 424-40 page 393
Splices, ground. elect. conductor 250-91a p.146
Splices, require box 300-15a p.168
Spontaneous ignition, dust 502-1 FPN p.518
Spotlight ports, projection from 540-10 p.624
Spray application pits 516-2a5 p.556
Spray area, portable lamps 516-3d p.562
Spray booth 516-2b2 p.556
Spread of Fire 300-21 page 171
Spreaders, cables in concrete 424-44c p.395
Sprinkler protect. panelboard 384-4a1 p.326
Stables, corrosive condition 300-6c FPN p.166
Stage cables, studios 400% 530-18a p.619
Stage lighting, demand factor T.530-19a p.620
Stage lights, branch circuit 20 amp 520-41 p.604
Stage lights, conduct. insulation 520-42 p.604
Stage pockets for receptacles 520-46 p.606
Stage set lighting cables 400% 530-18a p.619
Stage switchboard, metal hood 520-24 p.602
Stage, border lights 520-41 p.604
Stage, footlights 520-41 p.604
Stage, proscenium lights 520-41 p.604
Stage, receptacles 520-45 p.606
Stainless steel ground rods 250-83c2 page 146
Stair pressurization systems 517-34a4 p. 581
Stairs, clearance 230-9 p.91
Stairway chair lifts 620-1 p.706
Stanchions, aircraft hangars 513-6a p.543
Stand-alone system definition 690-2 p.777
Stand-by currents 665-61 ex. p.743
Standard classification branch circuit 210-3 p.53
Standard size overcurrent protection 240-6 p.110
Standby power, 60 seconds 701-11 p.798
Standby systems, battery 701-11a p.798
Standby systems, generator 701-11b p.798
Standpipes, floor receptacles 410-57d p. 371
Standstill, oper. torque motors 430-7c p. 419
Staring torque, modify 455-2 FPN p.483
Static and rotary converters 455-2 p.484
Static multipliers induction heat 665-60 p.742
Static, phase converter 455-2 p. 484
Static, phase converter 455-2 p.486
Stationary appliance, definition 550-2 p.631
Stationary motors 2 hp 430-109 ex.3 p.447
Stationary motors frame grd. 430-142 p.452
Steam, conductors in raceway 300-8 p.166
Steel plate, protection 300-4a1 page 160
Steel reinforcing bars 250-81c page 145
Steel siding, bonding 250-44 FPN p.133
Stepdown transformer 210-6c2 p.55

KEY WORD INDEX

Sticks, busway 364-12 page 298
Stiffen at temperatures 310-13 FPN page 182
Stock rooms, aircraft hangars 513-2d p.542
Storable pool, definition 680-4 p.756
Storable pools, GFCI 680-31 p.767
Storage batteries 480 p.490
Storage batteries, leads 640-9b p.731
Storage batteries, solar 690-71 p.784
Storage battery, emerg. system 700-12a p.793
Storage battery, standby system 701-11a p.798
Stored energy operator 710-21e5 p.812
Straight lines, underfloor raceway 354-8 p.287
Straight tubular lamps 410-73e ex.1 p.374
Straight voltage rating 240-85 p.119
Strain insulators open-conductors 225-13 p.86
Strain relief cable grips 364-8b4 page 298
Strain relief connector 370-23g1 page 308
Strain relief mobile home cord 550-5b p.633
Stranded conduct. on fixt. chains 410-28e p.365
Stranded type, outdoor lampholders 225-24 p.88
Strap of switch 380-10b page 323
Stress reduction means 710-6 p.807
Strip heaters 427-2 FPN page 409
Strip light 520-2 DEF p. 601
Strong chlorides 334-4 page 245
Structural applications 110-11 FPN2 p.39
Structural ceiling, dedicated space 384-4 p.326
Structural steel, bonding 250-80c p.144
Strut-type channel raceway 352-40 page 283
Studios, feeders 400% 530-18b p.619
Studs, bushing for cable 300-4b1 p.161
Subdivided load 48 amps 422-28f p. 386
Subdivided load, heat 48 amps 424-22b p.391
Subject to strain or phys. damage 410-30b p.366
Submersible pump, fountain 680-51b p.771
Submersible pumps 501-11 p.515
Submersion, occasional prolonged T430-91 p. 444
Substantial linoleum 354-3d page 286
Substantially increased 300-21 page 171
Subsurface enclosures 110-12b p.39
Suction systems 517-34a1 p.581
Suddenly moving parts 240-41b page 116
Sufficiently low impedance 250-51 p.135
Suffix LS 333-22 p.245
Suffix LS 352-21 p.282
Suffix LS nonmetallic wireway 362-27 p.294
Suffixes 310-11c p.181
Suitable balcony 430-132b p.452
Suitable covers on boxes 300-31 p.173
Suitable for wet locations 410-4a page 358
Suitable ground detectors 250-5b FPN p124
Suitable wiring methods 110-8 page 39
Sulfur hexafluoride breaker 230-204a page 105
Sulfur hexafluoride gas 325-21 page 273
Summation of the currents 430-24 ex.3 p.426
Sump pumps 517-34a2 p.581
Sunlight 362-16 (3) p.293
Sunlight effects 347-1 page 267
Sunlight resistance 553-7b p.685
Sunlight resistance 555-6 p.687
Sunlight resistant jacket 351-2 page 277
Sunlight-resistant 331-4 (9) page 241
Sunlight-resistant 340-5 page 255
Sunlight-resistant marking 310-11 d FPN p.181

Sunlight-resistant marking 402-9c FPN p.357
Sunlight-resistant marking 400-6b FPN p.347
Sunlight-resistant markings 347-17 FPN p.270
Sunlight-resistant solar cable 690-31b p.780
Supplemental electrode 250-81a ex. p.145
Supplementary overcurrent devices 240-10 p.110
Support fittings box fill 370-16b3 p. 305
Support hardware 300-6 page 165
Support of ceiling fans 422-18 page 383
Support of fixtures, PVC conduit 347-3b p.268
Support, strut-type channel raceway 352-47 p.284
Supported independently of box 410-16a p.363
Supporting screws for yokes 250-74 ex.2 p.141
Supports for ceiling fans 422-18 p.383
Supports for paddle fans 370-27c p.309
Supports for rigid conduit T.346-12 page 267
Supports for romex 336-18 p.250
Supports, vertical raceways T. 300-19a page 170
Suppressors, rectifier 665-24 p.740
Surface heating elements 60 amp 422-28b p. 385
Surface marking of conductors 310-11b1 p.180
Surface metal raceways, voltage 352-1 p. 280
Surface mining machinery 90-2b2 p. 21
Surface nonmetal. raceways, volts 352-22 p.282
Surface raceways 352-1 page 280
Surface raceways, splices & taps 352-7 p.281
Surface temperatures 547-3 p.628
Surface tracking 310-6 ex.a page 178
Surface-mounted incandescent 410-8d1 p.361
Surface-type snap switches 320-16 page 228
Surge arrester, 600v service 230-209 p.106
Surge arrester, grounding 280-25 page 158
Surge arrester, service 230-82 ex.4 p.101
Surge arresters definition 280-2 page 156
Surrounding metal by induction 300-20a p.171
Suspended ceiling 370-23c page 307
Suspended ceiling panels 300-23 p.173
Suspended ceilings 410-16c page 363
Suspension mount, chan. raceway 352-47b p.284
SWD, switches as breakers 240-83d p.119
Swimming pools 680 p.755
Swimming pools, bonding grid 680-22b page 764
Switch loop, cable 200-7 ex.2 p.50
Switch loops 380-2 ex. page 320
Switch, air 230-204 page 105
Switch, isolating 230-204d page 105
Switch, oil 230-204a page 105
Switch, oil definition 100 page 37
Switch, outlet, and tap devices 336-21 p. 250
Switch, vacuum 230-204a page 105
Switch, wall 210-70a p.66
Switchboard, as service, bonding 384-3c p.325
Switchboard, bottom clearance 384-10 p. 327
Switchboard, dead front 520-21 p.602
Switchboard, live parts 110-16a1 p.42
Switchboard, metal hood stage 520-24 p.602
Switchboard, pilot light 520-53g p.608
Switchboard, power lines 520-53o p.610
Switchboards 384 page 325
Switched lampholder 410-52 page 369
Switches, ungrounded wire 380-2a p.320
Switches, 3-way & 4-way connect. 380-2 p.320
Switches, 600 volt knife 380-16 page 324
Switches, AC snap 380-14a page 324

KEY WORD INDEX pg - 37

Switches, AC-DC snap 50% 380-14b2 p. 324
Switches, bull 530-15d p.618
Switches, canopy pull type 410-38c p.368
Switches, circuit breaker used as 380-11 p.323
Switches, CO/ALR snap 20 amp 380-14c p.324
Switches, double-throw knife 380-6b p. 321
Switches, enclosure 380-3 page 321
Switches, faceplates thickness 380-9 p. 322
Switches, flashers 600-6b1 p.691
Switches, foot shield 665-47b p.742
Switches, fused not in parallel 380-17 p.324
Switches, garbage disposal 422-25 p.385
Switches, gas pumps double-pole 514-5 p.549
Switches, grounded conductor 380-2b ex. p. 321
Switches, grounded cond. gas pump 514-5 p.549
Switches, highest position 380-8a page 322
Switches, horsepower rated 430-109 p.447
Switches, indicating off-on 380-7 p. 322
Switches, inductive load 380-14b2 p.324
Switches, isolat. Class I, Div.2 501-6b2 p.511
Switches, isolating 380-13a page 323
Switches, knife exposed blades 384-34 p. 330
Switches, manual adjustment 380-5 ex. p.321
Switches, motor control & disc. 430-111 p.449
Switches, mounting surface-type 380-10a p.323
Switches, mounting yoke 380-10b page 323
Switches, oil motor disconnect 430-111c p.449
Switches, pendant-surface-knife 380-3 ex.1 p.321
Switches, plaster ears 380-10b page 323
Switches, plates 380-9 p.322
Switches, pull type canopy 410-38c p.368
Switches, rated 347 volts 380-14d p. 324
Switches, simultaneously disc. 380-2b ex.1 p.321
Switches, single-throw knife 380-6a p. 321
Switches, surface-type snap 320-16 page 228
Switches, T-rated 380-14b3 p.324
Switches, thermostatically 424-20 page 390
Switches, time, flashers 6 380-5 ex. p. 321
Switches, voltage between 300v 380-8b p. 322
Switches, wet location 380-4 page 321
Switches, wire bending space 380-18 p.324
Switches. marking 380-15 p.324
Switching device, lampholder 410-48 page 369
Switching frames, telephone 220-3c ex.4 p. 71
Switching the grounded conductor 514-5a p.549
Synchronous generators 705-43 p.804
Systematic array 670-2 p.749

-T-

Table 1 Percent of fill conduit p.879
Table 10 PVC thermal expansion p.890
Table 5 insulated conductors area sq.in. p.883
Table 5a aluminum conduct. sq. in. area p.887
Table 8 DC resistance values p.888
Table 9 AC resistance values p.889
Table C1 Fixture wire percent of fill p.940
Table C1-12A conductors in conduit (fill) p.936
Table C1A compact conductors p.941
Tamper resistant receptacles 517-18c p.572
Tanks 225-19b p.88
Tap conductors 240-3d p.108
Tap conductors, fixtures 410-67c page 373
Taper per foot, conduit 500-2 p.493

Taper per foot, IMC conduit 345-8 page 261
Taper per foot, rigid conduit 346-7b page 264
Taps, cords not permitted 400-9 p.347
Taps, feeder not over 10' 240-21b p.112
Taps, feeder not over 25' 240-21c p.112
Taps, feeder over 25' 240-21e p.113
Taps, flat cable 363-10 p.295
Taps, motor feeder 430-28 p.427
Task illumination 517-33a p.580
Task illumination, definition 517-3 p.569
TC cable uses permitted 340-4 page 255
Technical equipment ground 530-72b page 623
Technology equipment Article 645 page 732
Telephone cond. pairs T.400-4 note 5 p.332
Telephone exchanges 220-3c ex.4 p.71
Telephone wires below power 800-10a1 p.853
Telescoping sections of raceway 250-77 p.142
Television equipment 810 p.865
Television studios 530 p.616
Temper.-actuated device 424-22d ex. c p.392
Temperature in excess 351-4b2 page 277
Temperature in fixtures 410-65 p.372
Temperature limitations 110-14c p.41
Temperature ratings 110-14c FPN p. 41
Temperature rise, busways 364-23 p.299
Temperature rise, motors 430-32a1 p.428
Temperature rise, reactors 470-18e p.490
Temperature, conductor limits 310-10 p.179
Temperature, hot to cold sealing 300-7a p.166
Temperature-limiting means 422-13 page 382
Temporary currents 250-21c page 125
Temporary lighting receptacles 305-4d p.174
Temporary power-light. permitted 305-3c p.174
Temporary wiring 90 days 305-3b p.174
Temporary wiring, approval 305-2b p. 174
Temporary wiring, emerg & tests 305-3c p. 174
Temporary wiring, GFCI 305-6a page 175
Temporary wiring, guarding 305-7 page 176
Temporary wiring, lamp protection 305-4f p.174
Temporary wiring, removal immed. 305-3d p.174
Temporary wiring, splices 305-4g page 175
Temporary wiring, tests 305-6b1 p.176
Ten sets of fuses 501-6b4 p.511
Tensile stress 501-4a p.505
Tension take-up device 400-8 ex. page 347
Tension will not be transmitted 400-10 p.347
Tension, fixture 410-28f page 365
Terminal fitting, raceway 300-5h p.165
Terminal housings, motors 430-12a p.421
Terminating seal MI cable 330-15 p.239
Termination fittings, MI cable 501-5 p.506
Test, cords 305-6b1 p.176
Test, emergency systems 700-4 p.789
Test, equip. grounding cond. 305-6b1 p.176
Test, ground fault on site 230-95c p.104
Test, heating cable 424-45 p.395
Test, lighting fixtures 410-45 p.369
Test, mobile home wiring 550-12 p.641
Test, receptacles 305-6b1 p.176
Test, recreational vehicle 551-60 p.663
Test, service GFCI 230-95c p.104
Test, standby systems 701-5 p.797
Test, temporary wiring 305-3c p.174
Theaters 520 p.600

KEY WORD INDEX

Theaters, border lights 520-44 p.605
Theaters, cord connectors 520-67 p.612
Theaters, curtain machines 520-48 p.606
Theaters, dimmers 520-25 p.602
Theaters, festoon wiring 520-65 p.611
Theaters, footlights 520-43 p.604
Theaters, metal hood switchboard 520-24 p.602
Theaters, portable power disb. box 520-62 p.611
Theaters, proscenium lights 520-44 p.605
Theaters, receptacles 520-45 p.606
Theaters, scenery 520-63 p.611
Theaters, special effects 520-66 p.612
Theaters, switchboards metal hood 520-24 p.602
Therapeutic pools 680-60 p.772
Therapeutic pools, GFCI 680-62a p.773
Thermal barrier, resistors 12 470-3 p.489
Thermal contraction, raceways 300-7b p.166
Thermal cutouts number/location T.430-37 p.432
Thermal devices for motors only 240-9 p.110
Thermal expansion 362-23 p.294
Thermal expansion or contraction 347-8 p.269
Thermal expansion or contraction 347-9 p.269
Thermal expansion, PVC 347-9 p.269
Thermal expansion, raceways 300-7b p.166
Thermal protection required 410-73 FPN p.374
Thermal protection, ballasts 410-73e page 374
Thermal protector 410-73 FPN p.374
Thermal protector definition 100 p. 35
Thermal shock, diffusers 410-4c2 p.359
Thermally noninsulating sand 424-41e p.394
Thermionic tubes Class I 501-3a p.503
Thermopl. insul., deformed 310-13 FPN p.182
Thermopl. insul. may be deform 402-3 FPN p.350
Thermopl. insul. may stiffen 402-3 FPN p.350
Thermoplastic insul., stiffen 310-13 FPN p. 182
Thermoset insulation T. 310-13 page 184
Thermosetting insulation, organs 650-5b p.726
Thermostatically controlled 424-20 page 390
Thermostats 424-20 page 390
Thickness of metal boxes 370-40b page 311
Thickness of sealing compound 501-5c3 p.508
Thorium 500-3b1 FPN p.499
Threaded bosses, bonding 250-72b p.140
Threaded couplings, bonding 250-72b p.140
Threaded enclosure 370-23d p.307
Threaded wrenchtight 370-23d page 307
Threaded, rigid conduit taper 346-7b p.264
Threadless concrete couplings 345-9a p.262
Threads, five fully engaged 501-4a p.505
Threads, taper per foot rigid steel 346-7b p.264
Threads, taper per foot, IMC conduit 345-8 p.261
Threads, taper per foot, NPT die 500-2 p.493
Three floors above grade, romex 336-5a1 p.248
Three-way switches 380-2a p.320
Thunderstorm days 800-30a FPN4(3) p.855
THW insulation, 90° C T.310-13 page 186
THW insulation, ballast compart. 410-31 p. 367
Tide level, marinas 555-3 FPN3 p.686
Tie ampacity, transformer 450-6a2 p.476
Tie conductors, transformer 450-6a4 ex. p.477
Tie wires, knobs #8 320-8 p.226
Tie wires, secure cables 300-19b3 p.170
Time delay restarting load 455-22 FPN p.486
Time delay, motor-comp. 440-54b p.468

Time switches, flashers 6" 380-5 ex. p. 321
Toasters 422-2 ex. page 379
Tool heads T.430-22a ex. p.425
Tools, portable grounded 250-45c p.134
Top shield, FCC cable 328-12a page 236
Torque motors 430-7c page 419
Torque requirements for controls 430-9c p. 420
Torque tightening 110-14 FPN p.40
Total hazard current 517-160b ex. p.598
Total hazard current, definition 517-3 p.567
Totally enclosed motors 501-8a p.512
Towel bars 680-41d4 ex. p.769
Toxicity characteristics, ENT 331-1 p.240
T.P. marking motors 430-7 (13) page 418
Tracer in braid, cords 400-22b p.348
Track as circuit conductor 610-21f4 p.703
Track conductors 410-105a page 378
Track lighting, voltages 410-105a p.378
Track load, fixtures 410-102 page 378
Trademark electrical equipment 110-21 p.44
Trailing cable 90-2b2 p.21
Tramrail track 610-21f4 p.703
Transfer equipment 230-83 page 102
Transfer equipment, emerg. systems 700-6 p.790
Transformer, askarel-insulated 450-25 p.480
Transformer, case not grounded 250-42ex.3 p.132
Transformer, doorways 450-43a p.482
Transformer, drainage 450-46 p.483
Transformer, dry-type 450-21a p.479
Transformer, dust 502-2a3 p.518
Transformer, fire pump 695-5b p.786
Transformer, grounded shield 426-32 page 407
Transformer, heat increase 450-3 FPN2 p.472
Transformer, individual 450-2 p.471
Transformer, isolation 427-27 page 412
Transformer, nameplate 450-11 p.478
Transformer, pool 680-5a p.757
Transformer, readily accessible 450-13 p.479
Transformer, stepdown 210-6c2 p.55
Transformer, zig-zag 450-5 p.475
Transformer-type rectifier 650-3 p.735
Transformers 450-2 p.471
Transformers, askarel-ins. 450-25 p.480
Transformers, auto sprinklers 450-43a ex. p.482
Transformers, carbon dioxide 450-43a ex. p.482
Transformers, chimney-flue 450-25 p.480
Transformers, circuits derived from 215-11 p.68
Transformers, combustible material 450-22 p.479
Transformers, current 250-122 ex. p.154
Transformers, dampers 450-45e p.483
Transformers, dielectric fluid 450-24 p.480
Transformers, donut-type 450-5a3 FPN p.476
Transformers, door locks 450-43c p.482
Transformers, drainage 450-46 p.483
Transformers, electric furnace 450-26 ex.3 p.481
Transformers, fire resistant 450-13ex.2 p.479
Transformers, frequency 450-11 p.478
Transformers, gases 450-25 p.480
Transformers, halon 450-43a ex. p.482
Transformers, impedance 450-11 p.478
Transformers, instrument 250-125 p.154
Transformers, insulating liquid 450-11 p.478
Transformers, located 12" 450-22 p.479
Transformers, location 450-13 p.479

KEY WORD INDEX   pg - 39

Transformers, marking 450-11 p.478
Transformers, nameplate 450-11 p.478
Transformers, outline lighting 600-21 p.692
Transformers, over 112 1/2 kva 450-21b p.479
Transformers, over 35,000v vault 450-21c p.479
Transformers, over 35kv in vault 450-24 p.479
Transformers, overcurrent prot. 450-6b p.477
Transformers, poorly vented 450-25 p.480
Transformers, potential 384-32 p.330
Transformers, pressure-relief vent 450-25 p.480
Transformers, readily accessible 450-13 p.479
Transformers, rooms fire resistant 450-21b p.479
Transformers, secondary ties O.C.P. 450-6b p.477
Transformers, sill height 450-43b p.482
Transformers, storage in vaults 450-48 p.483
Transformers, subtract.-connect. 450-5a3 p.476
Transformers, supply points tie 450-6a2 p.476
Transformers, T-connected 450-5 p.475
Transformers, temperature class 450-11 p.478
Transformers, tie ampacity 450-6a2 p.476
Transformers, vault 35kv 450-24 p.480
Transformers, vault door 450-43 p.482
Transformers, vault door vent size 450-45c p.483
Transformers, vault read. accessible 450-13 p.479
Transformers, vent openings 450-45a p.483
Transformers, ventilation 450-9 p.478
Transient motor 430-52c3 FPN p.435
Transition assembly 328-15 page 236
Transitory overvoltages 450-5c p.476
Translucent material 516-3c p.561
Transmission of noise or vibration 400-7a p.347
Transmission of noise, vibration 422-8c p. 380
Transmission of stresses 345-12 ex.1 page 262
Transmission of stresses 346-12 ex.1 page 266
Transposing cable size 370-28a2 page 310
Transverse raceway, header 356-1 p.288
Transverse raceway, header 358-2 p.288
Transversely routed, cablebus 365-6b page 301
Travel trailer definition 551-2 p.647
Traveling cables, lighting #14 620-12a1 p.710
Traveling cables, signaling #20 620-12a2 p.710
Traveling cables, supports 620-41 p.716
Tray cable, uses permitted 340-4 p. 255
Trees, light fixtures outdoors 410-16h p.363
Trees, live vegetation 225-26 p.89
Trench, cond. derating factor Note 8 ex.4 p.196
Trench, conductors run together 300-5i p.165
Trench-type flush raceway 354-6 ex. p. 287
Trench-type raceways 354-3c page 286
Trimming, PVC conduit 347-5 page 268
Trip coils number & location T.430-37 p.432
Trip free breakers 240-80 page 118
Trip setting 230-208 page 106
Trip-free circuit breakers 710-21a4b page 809
Triplen harmonic currents 310-4 FPN p. 177
Tripods, fixtures 410-16d page 363
Trolley frame, grounding 610-61 p.706
Trolley wires, ground return 110-19 p.44
Troubleshooting 690-13 FPN p.779
Troubleshooting testing 670-5 ex. p.751
Truck panels 230-204 ex. page 105
Tube keyers 665-24 p.767
Tube support, signs 600-41b p.694
Tubing for arms 410-38a page 368

Tubing, over 1000v marking 410-91 page 377
Tubular heaters 427-2 FPN page 409
Tungsten-filament lamp 380-14a2 page 324
Tunnel installations 710-51a p.816
Tunnel, equip. grd. conductor 710-54b p.817
Tunnels, discharge lamps 210-6d1b page 55
Turnbuckles 342-7b1 page 257
Turning vanes 424-59 FPN page 396
Turntables T.430-22a ex. p.425
Twisted or cabled, pendant cond. 410-27c p.365
Two appliances dedicated 210-8a2 ex.2 p.57
Two locknuts, bonding 250-76 ex.b p.142
Two-fer 520-2 DEF page 601
Two-fers, stage adapters 520-69 p.613
Two-wire DC systems 250-3a page 122

-U-

UF cable 339-1a page 253
UF cable, ampacity 339-5 page 254
UF cable, burial depth Table 300-5 p.163
UF cable, uses permitted 339-3a page 253
Ultimate insulation temp. T.400-5a note p.345
Ultimate insulation temperature T.520-44 p.605
Ultimate-trip current motors 430-126 page 451
Ultimate-trip setting motors 430-126 page 451
Unattended self-service stations 514-5c p.550
Unbroken lengths, wireways 362-9 p.292
Under chassis installations 551-10b5 page 648
Under raised floors, computer 645-5d p.733
Underfloor raceways 354-2 page 286
Underfloor raceways, conductors 354-5 p.286
Underfloor raceways, disc. outlet 354-7 p.287
Underfloor raceways, loop wiring 354-6 p.287
Underground block distribution 800-11b p.854
Underground cable under a bldg. 300-5c p.162
Underground excavators 710-41a p.815
Underground feeder 339-1a page 253
Underground installations 600v 300-37 p.173
Underground metal elbow 250-32 ex. p.131
Underground metal elbow 250-33 ex.4 p.132
Underground tanks 250-83b page 145
Underground wiring 514-8 ex.2 p.550
Underground, ducts 310-15d page 190
Underwater lights, pool 680-20a1 p.760
Underwater speakers, pool 680-23a p.764
Undue hazards DEF 100 p.29
Unequal division of current 310-4 FPN p.177
Unfinished accessory bldgs., GFCI 210-8a2 p.57
Unfinished basements, GFCI 210-8a5 page 57
Ungrounded cond. change in size 240-23 p.114
Ungrounded control circuits 685-14 p.775
Ungrounded systems 250-24b page 128
Unguarded live parts over 600v T.110-34e p.48
Unigrounded primary systems 280-24b1 p.158
Uninterruptible power supplies 645-11 p.735
Uniquely polarized 604-6b page 696
Unit equipment 700-12f p.794
Unit lighting load T.220-3b p.70
Unit switch, heaters 424-19c page 390
Unspliced grd. electrode conductor 250-81 p.144
Unswitched porcelain type 422-15a page 382
Unswitched type lampholders 410-62 page 359
Untrained, persons 90-1c p.21

KEY WORD INDEX

Unused openings, boxes 370-18 p.306
Unused openings, cabinets 373-4 p.313
Unused openings, equipment 110-12a p.39
Unwired portion, exist. dwelling 220-3d1 p.71
UPS equipment 645-11 ex.2 p.735
UPS systems 645-11 p.735
Upturned lugs 110-14a ex. p.40
Uranium 500-3b1 FPN p.497
Urban water-pipe areas 280-24a1 page 158
Usage characteristics 400-9 page 347
USE cable 338-1b page 251

-V-

Va per outlet, 180va 220-3c7 p.69
Vacuum cleaner cord T.400-4 page 340
Vacuum tubes 665-60 p.742
Vandalism 701-11 p.798
Vapor removal 300-22a page 171
Vapor seals, busway 364-25 ex. page 299
Vapor source 516-2a5 p.556
Vapor stop 516-2a5 p.556
Vapors and residues 516-3b p.561
Vapors of chlorine 334-4 page 245
Vapors, exposed to 110-11 page 39
Vaportight 410-4c2 p.359
Varnished cambric tapes T.400-4 note 3 p. 332
Varying duty definition 100 p.27
Varying electromagnetic field 665-2 p.740
Vault door vent size 450-45c p.483
Vault door, transformer 450-43 p.482
Vault, transformer 35kv 450-23 p.480
Vaults, film storage 530-51 p.621
Vaults, storage 450-48 p.483
Vehicle, door attached garage 210-70a p.66
Vehicle, electric charging 511-9 p.541
Vehicle, lanes lighting 511-7b p.541
Vehicle, mounted generator 250-6b p.124
Vehicle, self-propelled 511-1 p.539
Vehicle, washing areas 410-4a p.358
Vented cell flame arrestor 480-9a p.492
Ventilated, aircraft hangars 513-2d p.542
Ventilating equipment interlocked 516-2e p.561
Ventilation, battery rooms 480-8a p.492
Ventilation, computer/data room 645-5d3 p.734
Ventilation, motors 430-14a p.423
Ventilation, spray room 516-6d4 p.565
Ventilation, transformer 450-45a p.483
Ventilation openings, trans. vaults 450-45 p 483
Vertical clearance roof 225-19 ex.4 p.87
Vertical conductor support T.300-19a p. 170
Vertical Flame test 645-5d5 FPN p.734
Vertical metal poles cond. sup. 410-15b4 pa.362
Vertical or horizontal distances 250-42a p.132
Vertical raceways, cond. support T.300-19a p.170
Vertical risers, support 345-12 ex.2 p.262
Vertical shafts, spread of fire 300-21 p.171
Vertical tray flame test 725-71e FPN p.832
Vertically from top of bathtub rim 410-4d p.359
Vessel, definition 427-2 p.408
Vessel, stamped 424-72a page 397
Vibration, enduring 545-13 p.628
Video or radio frequency T.400-4 note 5 p.344
Viewing tables, lampholders 530-41 p.621

Visible gap, capacitors 460-24b2 p.488
Volatile disinfecting agents 517-60a2 p.590
Volatile flammable liquid DEF 100 p.35
Volatile flammable liquids 511-2 p.539
Volt-amps per square foot T.220-3b p.70
Voltage colors, heating cables 424-35 p. 393
Voltage converters 120v-low voltage 552-20b p.671
Voltage converters, vehicle 551-20b p.650
Voltage drop, ampacities 310-15 FPN p. 190
Voltage drop, branch circuit 210-19 FPN4 p. 58
Voltage drop, compensate cma 250-95 p.150
Voltage drop, feeders 215-2 FPN2 p.67
Voltage drop, phase converters 455-6 FPN p.484
Voltage drop, services 230-31c FPN page 94
Voltage drop, studio branch circuit 530-71d p. 623
Voltage marking on cables 725-71h FPN p.832
Voltage rating motors 430-83b p.442
Voltage rating not less than nominal 110-4 p.38
Voltage stresses, insulation 310-6 page 178
Voltage stresses, shielding 400-31b p.350
Voltage transformers 450-3c p.474
Voltage, 120/240 B. examples p.897
Voltage, contact with bodies 517-64a1 p.593
Voltage, DC aux. rectifiers 665-24 p.740
Voltage, limitations 300-2a page 159
Voltage, line to neutral 240-60a ex. p.117
Voltage, nominal 120/240 B. examples p.897
Voltages, circuit operates 110-4 page 38
Volts and amps, appliances 422-30a p. 387
Volts and watts, appliances 422-30a p. 387
Volume bounded 410-8a page 359
Vulcanized types 400-36 page 350

-W-

Wading pool, definition 680-4 p.756
Wall space, room dividers 210-52a p.63
Wall switch 210-70a p.66
Wall-mounted ovens, cord connect. 422-17a p.383
Wall-mounted ovens, grounding 250-60b p.138
Wall-mounted ovens, T.220-19 demands p. 75
Wallpaper, heating cables 424-42 p.395
Walls frequently washed 1/4" space 300-6c p.166
Walls, grounded 110-16a p.41
Warehouse demand factor T.220-11 page 72
Warning of derangement 700-7a p.790
Warning signs, Danger Hi-Voltage 230-203 p.105
WARNING TECHNICAL POWER 530-73 p.623
Wash cans, spray area 516-5d p.564
Washer/dryer Example 2b p.900
Washing machines, mobile home 550-11b3 p.640
Waste disposers, cord 422-8d1a page 381
Water accumulation, receptacles 410-57f p.371
Water heater, spa-hot tub 680-41h p.770
Water heaters, branch circuit 422-14b p. 382
Water heaters, pool 680-9 p.760
Water hoses 668-31 p.748
Water jets 680-41d4 ex. p.769
Water meters 250-81a page 144
Water pipe ground 250-81a p.144
Water pipes in vaults 450-47 p.483
Water spray 450-43a ex. p.482
Water, navigable wiring under 555-8 p.688
Waterfalls, stage lighting 520-66 p.612

… KEY WORD INDEX   pg - 41

Watt density, heating elements 426-20a p.405
Wattage, marking 110-21 p.44
Wave action 555-3 FPN3 p.686
Wave Shape DEF 100 p.31
Weakening the building structure 300-4a2 p.161
Weatherproof faceplate, recept. 410-57e p. 371
Weight, box support 50 pounds 410-16a p.363
Weight, fixture 6 pounds 410-15a p.362
Welders 630-1 p.725
Welders, ampacity 630-11 p.726
Welders, ampacity resistance 630-31 p.728
Welders, cable 630-41 p.729
Welders, duty-cycle 630-31b3 p.729
Welders, motor-gen arc 630-21 p.727
Welders, overcurrent 630-12a p.726
Welders, resistance 630-31 p.728
Welders, resistance O.C.P. 630-32a p.729
Welders, welds per hour 630-31b3 p.729
Well casings 250-43l p.133
Wellways, escalators 620-4 p.708
Wet bar sink 210-8a7 p.57
Wet location, cut-out box 373-2a p.313
Wet locations 300-6c page 166
Wet locations, boxes 370-15a p.303
Wet locations, fixtures 410-4a p.358
Wet locations, lampholders 410-49 p.369
Wet locations, polymeric braces 370-23b2 p. 307
Wet locations, receptacles 410-57 p.371
Wet locations, switch/breaker 380-4 p.321
Wet locations, tools & appl. 250-45d FPN p.134
Wet locations, wood braces 370-23b2 page 307
Wet-niche lighting fixture 680-25b1 p.765
Wet-pit 501-11 p.515
Wharfs 555-3 p.686
Wheelchair lifts 620-1 p.706
Whip, fixture metal flex. 250-91b ex.1 p.147
White, conductor 200-7 p.50
Windblown dust T.430-91 p.444
Winding with tape 400-10 FPN page 347
Windings the conductors energize 430-22 p.424
Windows, cords 400-11 p.348
Windows, service clearance 230-9 p.91
Wire bend. space, cab. wall opp. T.373-6b p.316
Wire bending space, cabinets T.373-6a p.314
Wire bending space, motors T.430-10b p.421
Wire binding screws 110-14a ex. p.40
Wire fill, conduit Tables C1-12A p. 941-1018
Wire mesh, animal confine. 547-8b p.630
Wire pulling compounds 517-160a6 p.597
Wire, bus, screw 250-79a page 142
Wire-to-wire connections T.430-12b p.422
Wired sections, heaters 424-12b page 389
Wiremold raceways 352-1 page 280
Wireways, 30 conductors 362-5 p. 291
Wireways, dead ends 362-10 p.292
Wireways, extensions from 362-11 p.292
Wireways, extensions thru walls 362-9 p.292
Wireways, sound recording 640-4 ex.a p.730
Wireways, splices & taps 75% 362-7 p.292
Wireways, supports 362-8 p.292
Wiring compartment 300-15d p.169
Wiring in ducts, plenums 300-22b page 172
Wiring methods, place of assembly 518-4p.599
Wiring, factory-installed internal 90-7 p.23

Wiring, under navigable water 555-8 p.688
Withdrawal of conductors 300-17 page 169
Within sight DEF 100 page 30
Within sight from DEF 100 page 30
Without permanent found. 550-2 p.631
Witness test emergency system 700-4a p.789
Wood braces mounting boxes 370-23b2 p.307
Wood braces, wet locations 370-23b2 p. 307
Wood, fiber or plastic 424-42 page 395
Wood joist temperature 331-3 FPN p. 240
Wood stud temperature 331-3 FPN p. 240
Wooden fixtures, line with metal 410-37 p.368
Wooden floors, open motors 430-14b p.423
Wooden plugs 110-13a page 40
Work applicator connections 665-64b p.743
Work areas 210-8a2 p.57
Work benches, cable over 342-7b3 p.257
Work light 520-2 DEF page 601
Work light 530-2 DEF page 617
Work lights, plugging boxes 530-18f p.619
Work space, access 110-33b page 46
Work surfaces counter tops 210-52c p.64
Working clearance 110-16a page 41
Workmanlike manner 110-12 page 39
Workmanlike manner 800-6 p.853
Workmanlike manner 820-6 p.871
Workspace is illuminated 110-16d FPN p.43
Wound-rotor code letter omitted 430-7a(8) p.418
Wound-rotor secondaries 430-32d p.430
Wrenchtight threaded couplings 250-72b p.140
Written record 230-95c page 104
Written record 700-4d p.790
Wye start, delta run connected 430-22a p. 425

-X-

X, superscript letter 90-3 p.22
X-ray equipment 660 p.736
X-ray equipment, O.C.P. 517-73b FPN p.595
X-ray feeders 660-6b p.737
X-ray, attachment plug 517-72c p.595
X-ray, conductors 660-6 p.737
X-ray, disconnect 660-5 p.737
X-ray, disconnect portable 517-72c p.595
X-ray, mobile definition 517-3 p.569
X-ray, mobile definition 660-2 p.737
X-ray, momentary rating 660-2 p.737
X-ray, nonmedical or nondental 660-1 p.736
X-ray, portable definition 660-2 p.737
X-ray, portable disconnect 517-72c p.595
X-ray, transportable 660-2 p.737
Xenon equipment 540-1 p.624
Xenon projectors, cond. size 540-13 p.625

-Y-

Yellow color, heating cable 120v 424-35 p.393
Yellow stripe, green conductor 250-59b p. 138
Yellow stripe, green conductor 310-12b p. 181
Yellow stripes 210-5b page 54
Yoke or strap, box 370-16b4 page 305
Yoke, multiwire circuit 210-4b page 54
Yokes 250-74 ex. 2 page 141

KEY WORD INDEX

-Z-

Z.P. marking 430-7 (13) page 418
Zener diode barriers 504-50a FPN p.534
Zig-zag transformers 450-5 p.475
Zinc 300-6a page 165
Zinc box thickness 370-40b p.311
Zinc galvanized rods 250-81c p.145
Zirconium 500-3b1 FPN p.497
Zone classifications Article 505 page 535
Zone directly over tub 410-4d page 359
Zone for fire ladders 225-19e page 88

"LEARN TO BE AN ELECTRICIAN"

1996 will mark the start of Tom Henry's NEW venture to actually train a person to become an electrician. With the purchase of 2 1/2 acres and the construction now completed for the Orlando headquarters, excitement is being generated throughout the country with interest in this new venture!

Starting with the toolbelt, safety, theory, practical wiring, Code, etc. Tom Henry will teach the student through his own personally designed **modules and animated videos** from point zero to an electrician that will not only know how, but *why*!

If you're a contractor that has electricians that have never had training, this will be an excellent opportunity to improve the quality and production of your work and have a person knowledgeable of the Code and able to communicate with the inspector.

This will be a correspondance course through the mail with any questions throughout the training course answered by mail, phone or fax. This training course will be offered throughout the *world*. For more information and start dates call today!

The most exciting training program ever developed! Now you'll be able to SEE the electron in orbit and SEE electricity through WATER ANALOGY in ACTION!!

1-800-642-2633

THE BIG SELLER

Now you can have a Code book just like the one Tom Henry uses!

THE "ULTIMATE" 1996 LOOSELEAF CODE BOOK FOR TAKING AN EXAM

It took over 3 1/2 hours but Tom Henry has personally **hi-lited** over *1300* answers to exam questions that have been asked on previous electrical examinations.

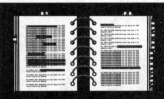

68 TABS ARE INSTALLED FOR YOU!

DON'T DELAY
ORDER YOURS TODAY!
1-800-642-2633

COLOR
YELLOW = JOURNEYMAN
PINK = MASTER
ORANGE = BOTH Journeyman & Master

The Journeyman exam answers are hi-lited with a "yellow" marker, the answers for the Master exams are marked with a "pink" marker. Questions which have been asked on both the Journeyman and Master exams are marked "orange" in color.

The "ULTIMATE" Looseleaf Code book also includes **Tom Henry's** BIG SELLER the "KEY WORD INDEX","REMINDERS FOR THE ELECTRICIAN", 68 "CODE TABS" (**installed for you**) and the popular "FORMULA INSERT

PRE-EXAMS

Order your *complete* Journeyman or Master pre-exam today!

Why take an actual exam, spend all that money, often lose a day of work, when you're not sure if you'll pass?

NOW THE ANXIETY CAN BE REMOVED FROM EXAM TAKING!

Tom Henry has now computer designed actual electrical exams with the percentage of questions from different catagories based just as the exams are. The level of mastery is perfected in each exam. A perfect way to tell if you're ready for the exam.

After completing your *complete* electrical exam you will mail your answer sheet back to Tom Henry and he will personally computer grade it and a grade evaluation sheet will be returned in the same day's mail. You'll know right away your strong and weak areas as the evaluation will point out each in detail so you'll know exactly where you stand with Tom Henry's personal evaluation.

| EXAMS WILL DIFFER FROM MONTH TO MONTH |

ITEM # 280 - $20.00   JOURNEYMAN COMPLETE 6 HOUR EXAM
50 CLOSED BOOK QUESTIONS   50 OPEN BOOK QUESTIONS
30 CALCULATIONS

ITEM #281 - $20.00   MASTER COMPLETE 6 HOUR EXAM
50 CLOSED BOOK QUESTIONS   70 OPEN BOOK QUESTIONS
30 CALCULATIONS

 Call 1-800-642-2633 Today!

The troublesome CLOSED BOOK is now made easy as Tom Henry goes over each answer in DETAIL on audio tape.

3 CLOSED BOOK exams with **150** questions selected by Tom Henry to help the exam applicant learn this part of the electrical examination. Each CLOSED BOOK EXAM contains 50 questions. A ONE HOUR tape narrated by Tom Henry gives, in full detail, the answer to each question on each exam. **A total of 3 hours of audio!**

You will learn how to memorize and store items in your mind. These exams can be worked over and over again. **The 3 audio tapes can be listened to as you drive in your truck or as you sit at home.** The student will memorize burial depths, definitions, service clearance heights, etc.

With **ITEM #192** you will receive a workbook containing 3 closed book exams plus 3 audio tapes by *Tom Henry*. Price $29.00 (plus tax for Florida residents only) include $5 for shipping and handling.

THREE HOURS OF TAPES FOR MEMORIZATION - THE KEY TO CLOSED BOOK